TECHNICAL
REPORT

The Impact of Air Quality on Hospital Spending

John A. Romley, Andrew Hackbarth,
Dana P. Goldman

Sponsored by the The William and Flora Hewlett Foundation

HEALTH

This work was sponsored by the William and Flora Hewlett Foundation. The research was conducted in RAND Health, a division of the RAND Corporation.

Library of Congress Cataloging-in-Publication Data

Romley, John A.
The impact of air quality on hospital spending / John A. Romley, Andrew Hackbarth, Dana P. Goldman.
p. cm.
Includes bibliographical references.
ISBN 978-0-8330-4929-2 (pbk. : alk. paper)
1. Air—Pollution—Health aspects—California. 2. Air—Pollution—Economic aspects—California.
3. Hospitals—California—Cost of operation. 4. Medical care, Cost of—California. I. Hackbarth, Andrew.
II. Goldman, Dana P. (Dana Paul), 1966– III. Rand Corporation. IV. Title.
[DNLM: 1. Air Pollution—adverse effects. 2. Air Pollution—economics. 3. Economics, Hospital.
WA 754 R765i 2010]

RA576.6.C2R66 2010
363.739'209794—dc22

2010000853

Published 2010 by the RAND Corporation
1776 Main Street, P.O. Box 2138, Santa Monica, CA 90407-2138
1200 South Hayes Street, Arlington, VA 22202-5050
4570 Fifth Avenue, Suite 600, Pittsburgh, PA 15213-2665
RAND URL: http://www.rand.org/
To order RAND documents or to obtain additional information, contact
Distribution Services: Telephone: (310) 451-7002;
Fax: (310) 451-6915; Email: order@rand.org

Preface

Air pollution is detrimental to human health, with adverse effects that include emergency room visits for asthma and hospitalizations for respiratory and cardiovascular causes. Little is known about the financing of such pollution-related care. This care may impose a significant burden on insurance companies and employers, and also on public programs such as Medicare and Medicaid.

This study determined how much failing to meet federal air quality standards cost various purchasers/payers of hospital care in California over 2005–2007. This report describes the analytical approach and discusses the study findings and their implications. Its contents should be of interest to health care payers, policymakers, health and environmental researchers, and the broader public.

This work was sponsored by the William and Flora Hewlett Foundation. The research was conducted in RAND Health, a division of the RAND Corporation. A profile of RAND Health, abstracts of its publications, and ordering information can be found at www.rand.org/health.

Contents

Figures

Tables

Summary

Air pollution is detrimental to human health, with adverse effects that range from restrictions in physical activity, to emergency room (ER) visits for asthma and hospitalizations for respiratory and cardiovascular causes, to premature mortality. The economic costs of such effects are substantial.

Little is known, however, about the financing of pollution-related health care. If such care imposes a significant burden on insurance companies and employers, they would have substantial stakes in improving air quality. Reduced medical spending could also benefit public programs such as Medicare and Medicaid.

The primary objective of this study was to determine pollution-related medical spending by private health insurers as well as by public purchasers such as Medicare. The study focused exclusively on hospital spending, though doctors' visits and other medical care also result from air pollution. The study did not address broader impacts of air pollution on health, which are important but better understood. (Table S.1 lists more health endpoints associated with ozone and PM2.5, identified in U.S. Environmental Protection Agency [2008c], but is not exhaustive.) The cost of air quality improvement is also important, but outside our scope.

To pursue our objective, we quantified the hospital spending incurred by health care purchasers/payers from 2005 to 2007 that is attributable to California not meeting clean air standards. Millions of people were exposed to significant air pollution during this period. In addition, the state of California collects and discloses appropriate clinical and financial data on hospital care, in particular, data on spending by payers for pollution-related admissions for cardiovascular and respiratory causes, and ER visits for asthma. As the report describes in detail, we used epidemiological studies and actual pollution patterns to determine how meeting federal air quality standards would affect the number of acute health events requiring hospital care. We used actual patterns of hospital care to determine the potential reductions in care delivered at specific hospitals. Finally, we used actual spending patterns to quantify the cost, and therefore the potential spending reductions, for different types of payers.

Table S.2 summarizes our overall results. Meeting federal clean air standards would have prevented an estimated 29,808 hospital admissions and ER visits throughout California over 2005–2007. (To prevent double counting, hospital admissions are defined to include hospital encounters that began in the ER but that led to an admission.) Nearly three-quarters of the potentially prevented events are attributable to reductions in ambient levels of fine particulate matter, that is, particulate matter with an aerodynamic diameter of less than or equal to 2.5 micrometers, which we abbreviate as PM2.5. The rest of the prevented events are attributable to reductions in ozone.

Failing to meet federal clean air standards cost health care purchasers/payers $193,100,184 for hospital care alone. In other words, improved air quality would have reduced total spending

Table S.1
Known and Quantified Health Endpoints Associated with PM2.5 and Ozone

PM2.5 health effects	Premature mortality
	Chronic and acute bronchitis
	Respiratory hospital admissions
	Cardiovascular hospital admissions
	ER visits for asthma
	Heart attacks (myocardial infarction)
	Lower and upper respiratory illness
	Minor restricted-activity days
	Work loss days
	Asthma exacerbations (asthmatic population)
	Respiratory symptoms (asthmatic population)
	Infant mortality
Ozone health effects	Premature mortality: short-term exposures
	Respiratory hospital admissions
	ER visits for asthma
	Minor restricted-activity days
	School loss days
	Asthma attacks
	Acute respiratory symptoms

NOTE: Shading indicates endpoints included in this study.

Table S.2
Air Pollution–Related Hospital Events, Spending, and Hospital Charges in California over 2005–2007 Caused by Failure to Meet Federal PM2.5 and Ozone Standards, by Pollutant, Endpoint, and Population

Pollutant	Endpoint	Population	Events	Spending	Hospital Charges
Ozone	Acute bronchitis, pneumonia, or COPD admission	All ages	6,056	$56,500,000	$226,000,000
PM2.5	Pneumonia admission	65 and older	2,517	$27,700,000	$123,000,000
PM2.5	COPD admission	65 and older	652	$5,634,450	$24,800,000
PM2.5	COPD admission excl. asthma	Age 18–64	306	$2,721,382	$10,900,000
PM2.5	Asthma admission	64 and younger	940	$5,575,469	$20,100,000
PM2.5	Any cardiovascular admission	65 and older	3,256	$47,700,000	$205,000,000
PM2.5	Any cardiovascular admission	Age 18–64	1,864	$35,100,000	$120,000,000
Ozone	Asthma ER visit	All ages	2,027	$1,768,883	$5,271,011
PM2.5	Asthma ER visit	17 and younger	12,190	$10,400,000	$31,700,000
Total			**29,808**	**$193,100,184**	**$766,771,011**

on hospital care by $193,100,184 in total. Table S.3 reports cost by type of payer. Medicare, the federal program that primarily covers the elderly, spent $103,600,000 on air pollution–related hospital care during 2005–2007. Medicaid (Medi-Cal in California), the federal-state program that covers low-income people, spent $27,292,199. Private health insurers (that is, third-party payers) spent about $55,879,780 on hospital care.

These results suggest that the stakeholders of public programs may benefit substantially from meeting federal clean air standards. Private health insurers and employers (who contribute to employee health insurance premiums) may also have sizable stakes in improved air quality.

We also determined the impact of poor air quality at specific hospitals. Five hospitals are presented here as "case studies": Riverside Community Hospital, St. Agnes Medical Center, St. Francis Medical Center, Stanford University Hospital, and University of California–Davis Medical Center.

These case studies are a diverse group. We reviewed and qualitatively selected hospitals according to the following criteria: the scale of potential prevented events and spending reductions; geographic region; and payer and patient mix.

Figure S.1 shows the number of events by patient zip code. These events are concentrated in the San Joaquin Valley and South Coast air basins. St. Agnes is located in the former, while Riverside Community and St. Francis are located in the latter. PM2.5 and ozone levels in these areas substantially exceed federal standards. A sizable number of events originate in and near Sacramento, where the UC Davis Medical Center is located.

Stanford University Hospital is located in the San Francisco metropolitan area. Moreover, as Table S.4 shows, private insurers were expected to pay most of the bill for 46% of Stanford University Hospital's patients, versus 31% for California as a whole. At the other extreme, private payers paid for only 14% of patients at St. Francis. Medi-Cal paid for 59% of patients, compared with a state average of 22%. Among the case study hospitals, the Medicare share was highest at St. Agnes (50%) and lowest at St. Francis (21%).

The racial composition of patients varied substantially across hospitals. Slightly more than three-quarters of patients were white at Stanford University Hospital, compared with 2% at St. Francis. African-Americans were 20% of the patient population at St. Francis, compared with

Table S.3
Events, Spending, and Hospital Charges in California over 2005–2007 Caused by Failure to Meet Federal PM2.5 and Ozone Standards, by Payer Type

Payer	Reduction in Events	% of Total Event Reduction	Reduction in Spending	% of Total Spending Reduction	Reduction in Hospital Charges
Medicare	9,247	31.02	$103,600,000	53.60	$463,000,000
Medi-Cal	8,982	30.13	$27,292,199	14.14	$126,000,000
County indigent	335	1.12	$1,071,967	0.55	$7,612,133
Total public	18,564	62.28	$131,964,166	68.29	$596,612,133
Total private third-party	9,029	30.29	$55,879,780	28.90	$149,954,889
Total all other	2,216	7.43	$5,443,008	2.82	$20,919,389

NOTE: Medi-Cal is the name for California's Medicaid program.

Figure S.1
Pollution-Related Hospital Events throughout California over 2005–2007, by Patient Zip Code

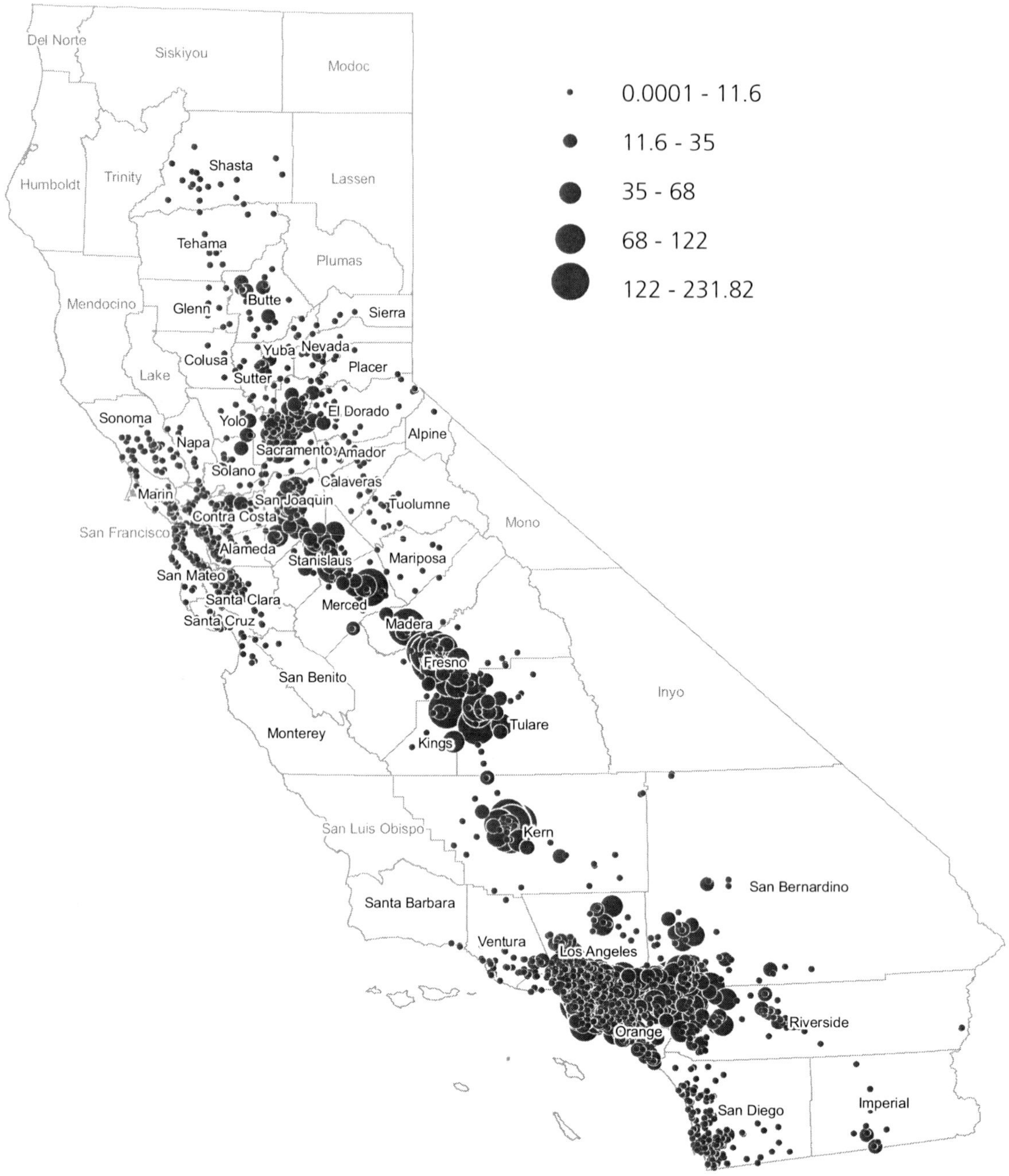

a statewide average of 7%. The proportion of Hispanics patients was well above average at St. Francis (77%) and at Riverside Community Hospital (38%).

The economic status of patients also varied widely. Statewide, 15% of patients have incomes below the federal poverty level. But at St. Francis, more than one-quarter of patients were poor; at Stanford University Hospital, fewer than 10% of patients were poor.

Figures S.2 through S.11 show the number of air pollution–related events at each of the five case-study hospitals:

Table S.4
Characteristics of Case Study Hospitals, 2005–2007

Hospital	Riverside Community Hospital	St. Agnes Medical Center	St. Francis Medical Center	Stanford University Hospital	UC Davis Medical Center	All California Hospitals
Summary information						
City	Riverside	Fresno	Lynwood	Stanford	Sacramento	—
County	Riverside	Fresno	Los Angeles	Santa Clara	Sacramento	—
Annual discharges	18,903	24,396	22,841	22,788	29,282	7,248
Staffed beds	345	406	384	454	550	175
Teaching hospital	No	No	No	Yes	Yes	—
Discharges, by payer (%)						
Private third-party	37	30	14	46	35	31
Medicare	36	50	21	38	24	37
Medi-Cal	22	18	59	9	29	22
Other	5	2	7	7	13	10
Patient race/ethnicity (%)						
White	51	72	2	78	50	62
Black	7	4	20	5	12	7
Hispanic	38	21	77	7	18	24
Asian or Pacific Islander	1	3	0	10	5	5
American Indian	0	0	0	0	0	0
Other	3	0	1	0	15	2
Patient economic status, by income as percentage of Federal Poverty Level						
0–100% FPL	15	20	27	9	16	15
> 100% FPL	85	80	73	91	84	85

NOTES: Medi-Cal is the name for California's Medicaid program. See Table 4.4 for detailed payer types. Racial groupings include non-Hispanic persons of single race.

At **Riverside Community Hospital**, 329 hospital admissions and ER visits would have been prevented had federal standards for PM2.5 and ozone been met during 2005–2007 (Figure S.2). Private health insurers paid most of the bill for almost half (149) of these patients. Medicare was the next most frequent payer for these preventable events. Overall, spending was $2,015,880 (Figure S.3). Medicare spent about $1,140,060, as these patients were relatively likely to have costly hospital stays, rather than ER visits. Private insurers spent $708,700.

At **St. Agnes Medical Center** in Fresno, failing to meet federal air standards had even greater effects: 384 hospital admissions/ER visits occurred (Figure S.4) and $2,976,936 was spent (Figure S.5). More than half of these events (208), totaling $1,913,116, were paid for primarily by Medicare, consistent with its above-average importance at this hospital.

At **St. Francis Medical Center** in Lynnwood (south of Los Angeles), 295 hospital admissions and ER visits occurred (Figure. S.6). Medi-Cal was the primary payer for more than half of these events (156). The next most frequent payer, Medicare, had one-third as many events (51). Nevertheless, Medicare spent $716,979, partly because Medi-Cal tends to pay less for hospital care. For example, Medi-Cal spent $9,482 on average for pneumonia admissions for those 65 and older, compared with $10,882 for Medicare. Overall, failing to meet clean air standards led to $1,220,595 in spending at St. Francis (Figure. S.7).

At **Stanford University Hospital**, 30 hospital admissions and ER visits occurred (Figure S.8), costing $534,855 (Figure. S.9). Figure S.1 shows that fewer events occurred in the San Francisco metro area than in other parts of the state.

At **UC Davis Medical Center** in Sacramento, our final case study, 182 events occurred (Figure. S.10), and spending totaled $1,882,412 (Figure S.11). Medi-Cal was the most frequent payer (81) for these preventable events, while Medicare would have experienced the largest spending reduction ($855,499).

These case studies underscore that health care payers could enjoy substantial reductions in hospital spending from improved air quality. The payers who benefit the most vary substantially across hospitals and communities.

Figure S.2
Air Pollution–Related Hospital Events at Riverside Community Hospital over 2005–2007, by Payer

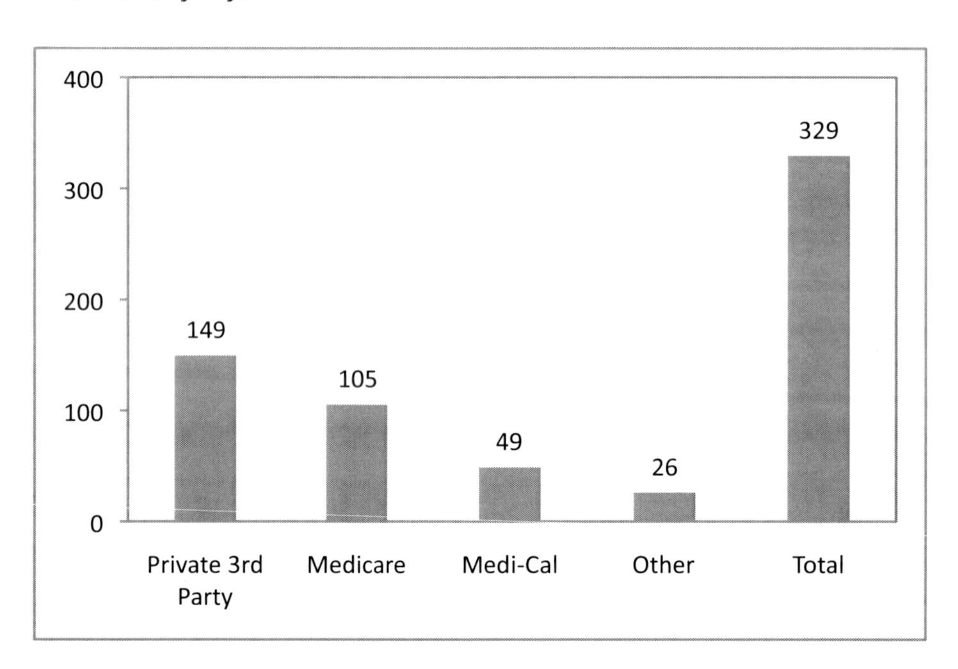

Figure S.3
Air Pollution–Related Hospital Spending at Riverside Community Hospital over 2005–2007, by Payer

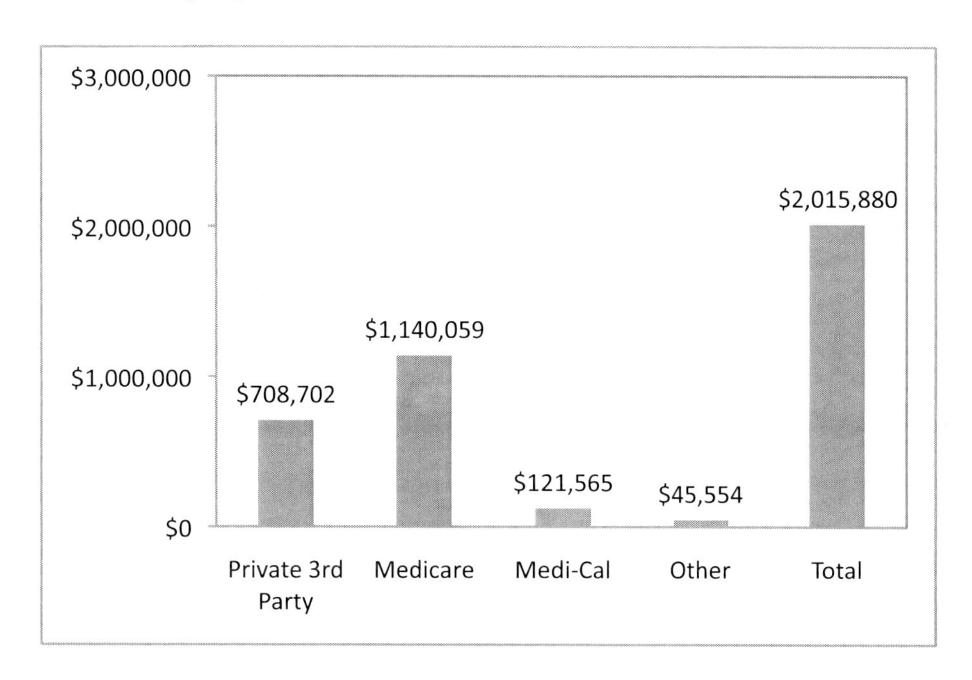

Figure S.4
Air Pollution–Related Hospital Events at St. Agnes Medical Center over 2005–2007, by Payer

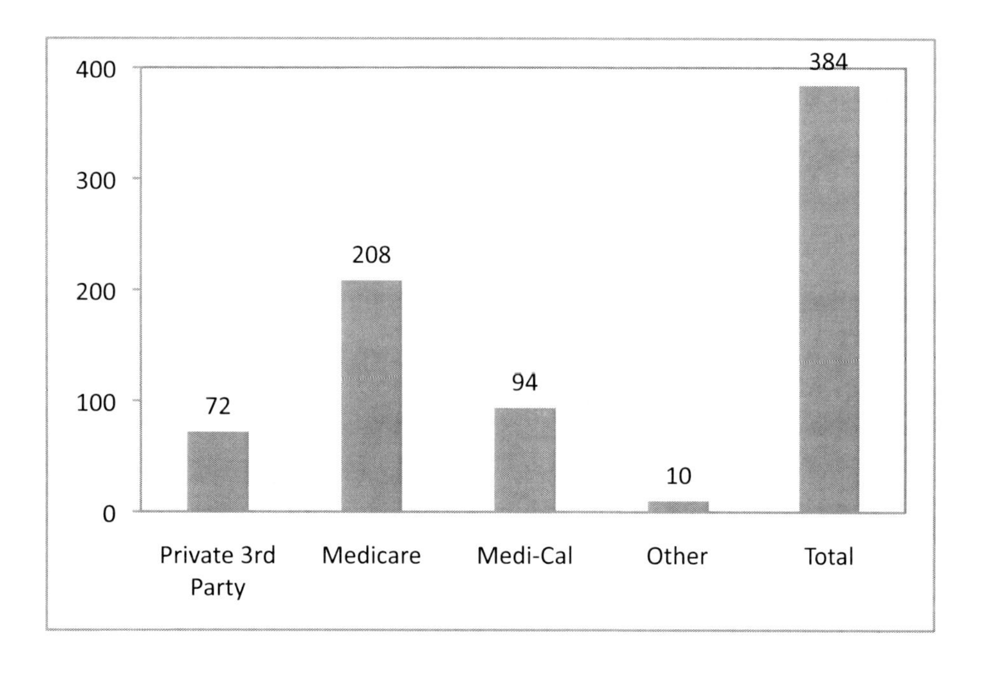

Figure S.5
Air Pollution–Related Hospital Spending at St. Agnes Medical Center over 2005–2007, by Payer

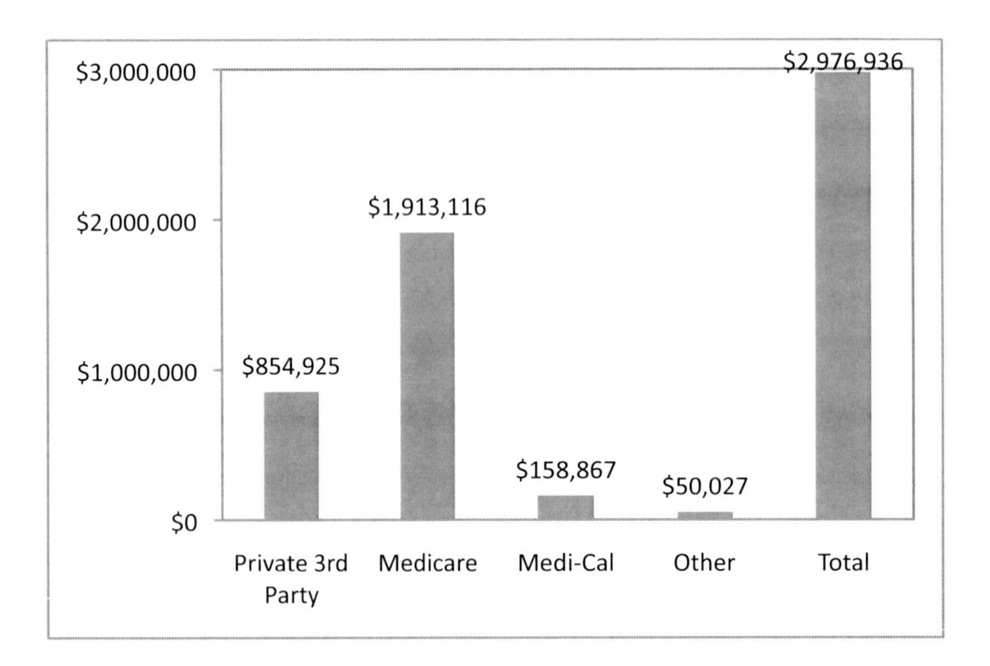

Figure S.6
Air Pollution–Related Hospital Events at St. Francis Medical Center over 2005–2007, by Payer

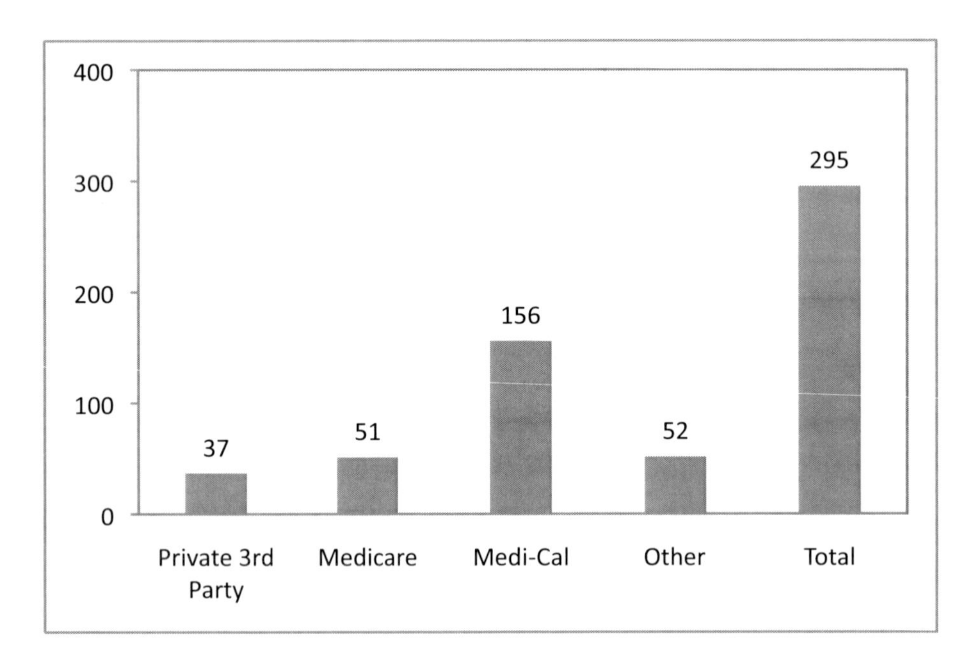

Figure S.7
Air Pollution–Related Hospital Spending at St. Francis Medical Center over 2005–2007, by Payer

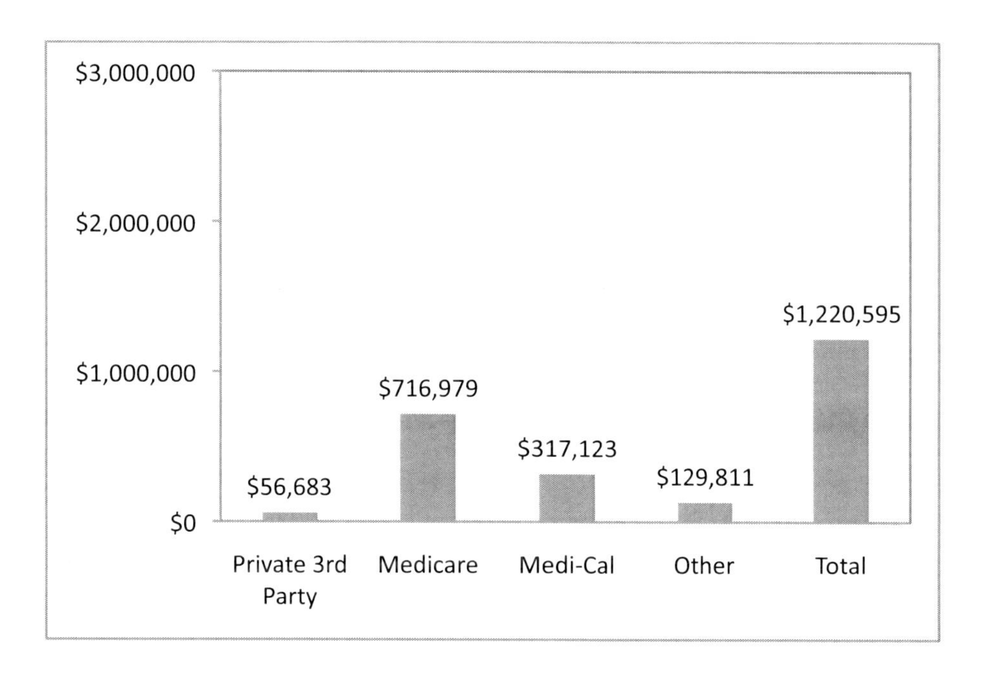

Figure S.8
Air Pollution–Related Hospital Events at Stanford University Hospital over 2005–2007, by Payer

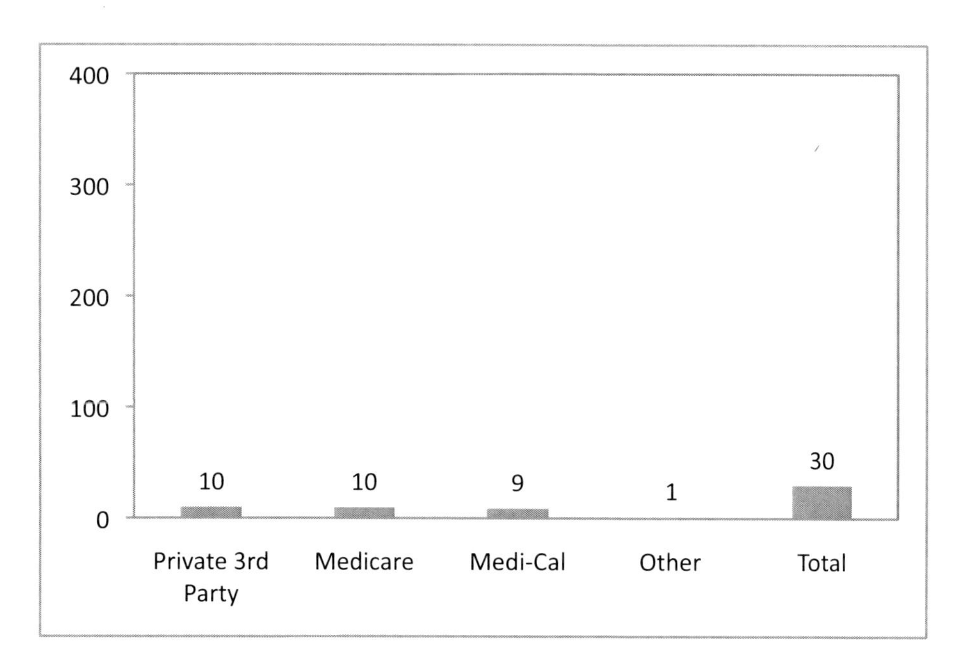

Figure S.9
Air Pollution–Related Hospital Spending at Stanford University Hospital over 2005–2007, by Payer

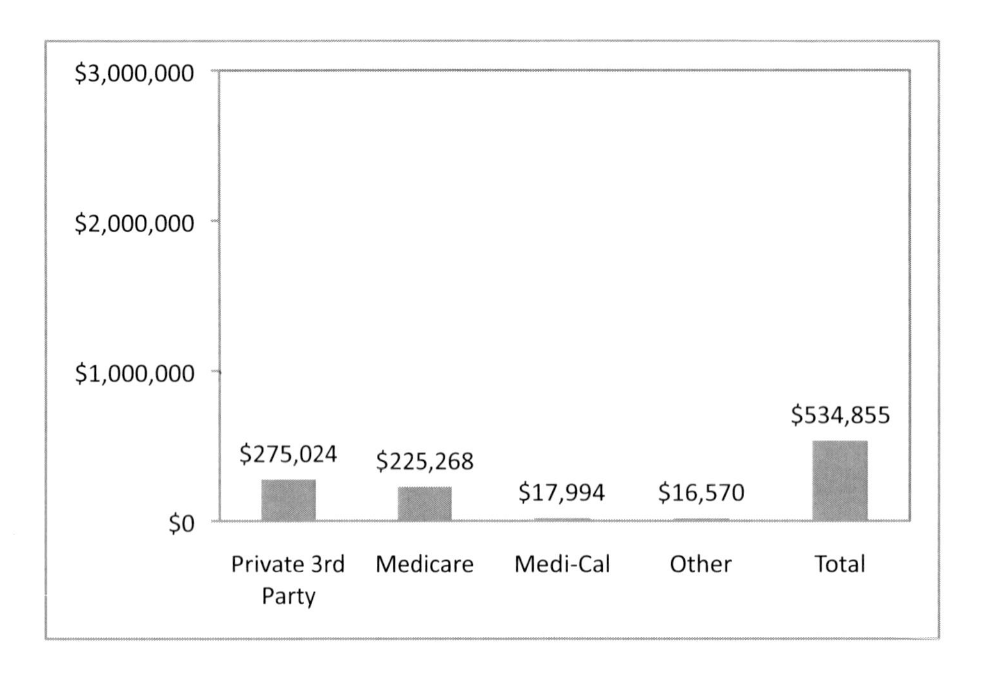

Figure S.10
Air Pollution–Related Hospital Events at UC Davis Medical Center over 2005–2007, by Payer

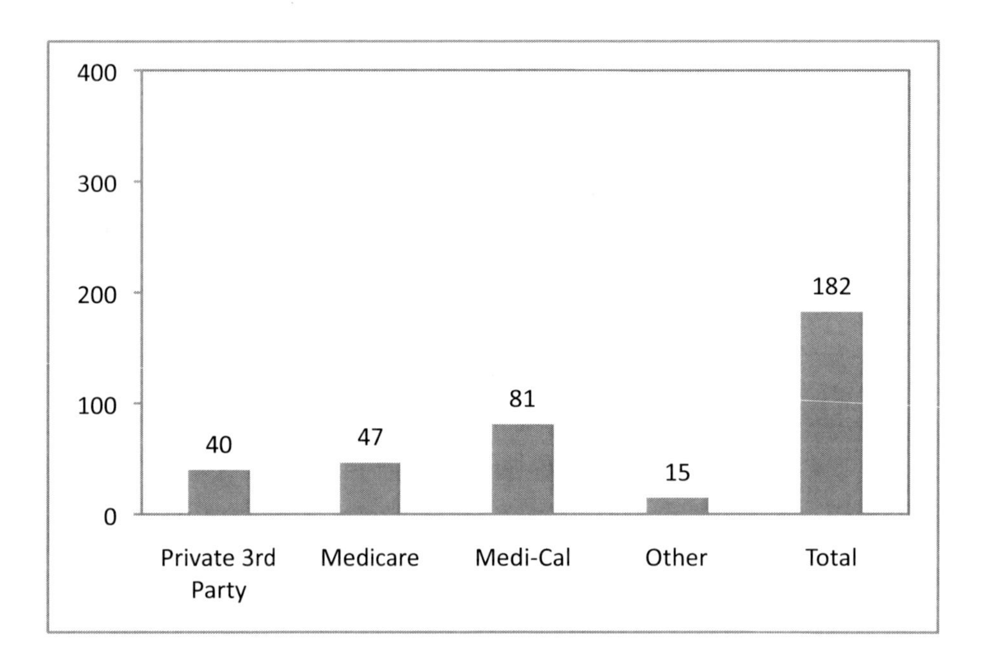

Figure S.11
Air Pollution–Related Hospital Spending at UC Davis Medical Center over 2005–2007, by Payer

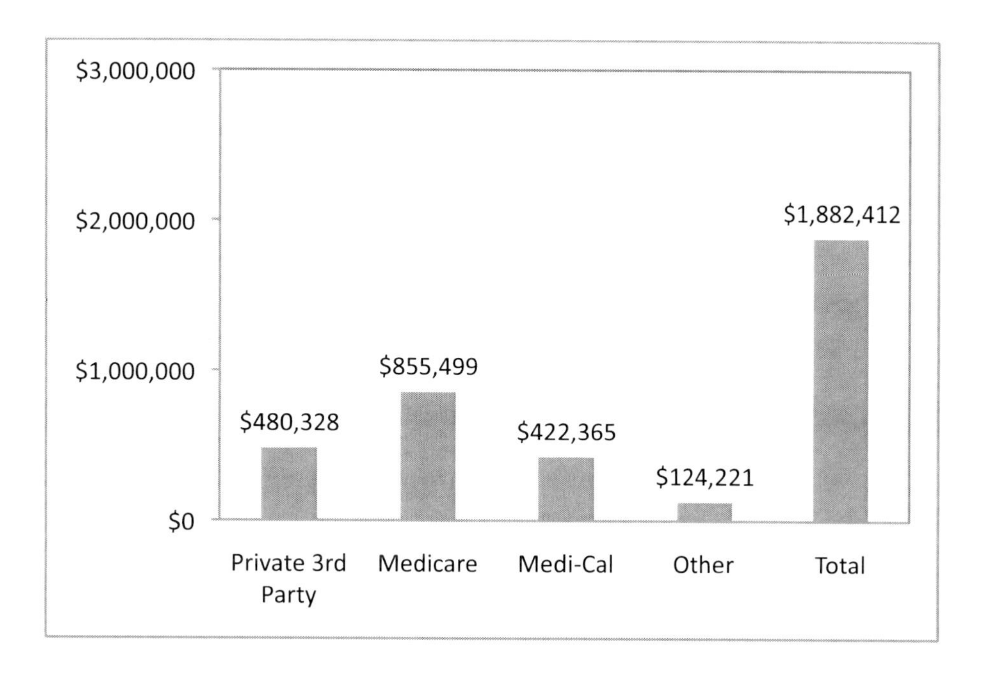

Acknowledgments

We are greatly indebted to program officer Danielle Deane for her tireless support of this study. Lloyd Dixon and Bart Ostro provided very valuable comments on this report. Mary Vaiana and Michelle Mulkey offered helpful assistance with communications. Nancy Good helped with the document's formatting.

Abbreviations

BenMAP	Environmental Benefits Mapping and Analysis Program
COPD	chronic obstructive pulmonary disease
EPA	U.S. Environmental Protection Agency
ER	emergency room
ICD-9	International Statistical Classification of Diseases and Related Health Problems
PM2.5	particulate matter with an aerodynamic diameter of less than or equal to 2.5 micrometers

CHAPTER ONE

Introduction

It is well established that air pollution harms human health. Adverse effects range from minor restrictions in activity to emergency room (ER) visits for asthma, hospitalizations for respiratory and cardiovascular causes, and premature mortality (U.S. Environmental Protection Agency [EPA] 1999a). Health effects such as these can reduce the well-being of society. Much is known about the overall magnitude of harm from air pollution in economic terms; for example, the cost of pollution-related hospitalizations has been analyzed extensively (EPA 1999b).

Yet little is known about the financing of pollution-related medical care. Such care may impose a significant burden on insurance companies and employers, who would then have sizable stakes in improved air quality. Similarly, reduced medical spending could also be of substantial benefit to public programs such as Medicare and Medicaid.

The primary objective of this study is to assess the cost of dirty air, exclusively as it relates to hospital care, for health care purchasers. Such an assessment requires reliable information about the spending of different purchasers/payers on pollution-related medical care, ideally within a large and important portion of the United States.

We analyzed hospital admissions for respiratory and cardiovascular causes, and ER visits for asthma, throughout California from 2005–2007. The state of California collects and discloses appropriate clinical and financial data on hospital care. In the analysis, we first used peer-reviewed epidemiological studies and actual pollution patterns to measure the impact that meeting or failing to meet clean-air standards for particulate matter and ozone has on the number of acute health events requiring hospital care throughout the state. We then used actual patterns of hospital care to determine the care delivered at specific hospitals. Finally, we used actual spending patterns to quantify the cost—that is, what health care purchasers spent because California failed to meet clean-air standards—for different types of payers.

The study did not address the costs to purchasers of pollution-related medical care delivered outside of hospitals (for example, in doctors' offices). We also did not address broader impacts of air pollution on health, such as restrictions on activity or premature mortality. These effects are important but well understood. We also did not address the costs of meeting clean-air standards. These costs are relevant to the policy debate, but beyond the scope of the present study.

The rest of this report is organized as follows: Chapter Two reviews relevant background information. Chapter Three describes our data and analytical methods. Chapter Four presents the results, and Chapter Five concludes.

CHAPTER TWO

Background

Fuchs and Frank (2002) analyzed spending on health care related to air pollution using claims data on elderly whites in 183 metro areas throughout the United States. These authors found, after accounting for confounding factors such as smoking, that a 10 microgram per cubic meter reduction in average annual levels of particulate matter with an aerodynamic diameter of less than or equal to 10 micrometers decreases average outpatient spending by $100, and inpatient (that is, hospital) spending by $77, for these Medicare patients.[1]

Yet Medicare primarily covers elderly Americans, and thus only a fraction of the U.S. population. As Table 2.1 shows, U.S. health spending totaled $2,241 billion in 2007 (Hartman, Martin, et al. 2009). Medicare accounted for $431 billion, or 19.2% of the total. Medicaid, the state and federal program for low-income persons, accounted for another $329 billion. In terms of private funding, health insurance spent $775 billion, or 34.6% of the total, while Americans spent $269 billion out of pocket.

In order to deepen our understanding of the financing of air pollution–related care, we analyzed spending on hospital care in California in recent years.

Rich information on hospital spending by type of purchaser/payer is available for California. For medical spending, actual payment is of interest. For hospital care, the distinction

Table 2.1
National Health Spending in 2007, by Source of Funds

Payer/Purchaser	Health Spending (billions)	Percentage of Spending
Medicare	$431	19.2
Medicaid	$329	14.7
Other public	$275	12.3
Personal out-of-pocket payments	$269	12.0
Private health insurance	$775	34.6
Other private funds	$162	7.2
	$2,241	**100.0**

Source: Hartman, Martin, et al. (2009).

[1] Chestnut, Thayer, et al. (2006) estimate medical costs (if not spending) after hospitalizations for respiratory and cardiovascular causes. As discussed below, pollution can lead to such hospitalizations. Post-hospitalization costs ranged from 13–24% of hospital spending, as estimated based on the average discount off charges in California hospitals over 2005–2007 (discussed below in this chapter, and also in Chapter Three). For additional information on medical costs for pollution-related conditions, see U.S. Environmental Protection Agency (1999b).

between actual payment/spending and billed charges is critical. Hospital bills list charges at "full established rates" that do not reflect discounts negotiated with payers or other deductions from "gross revenue." For California over 2005–2007, the average discount off charges was 74.8% (California Office of Statewide Health Planning and Development 2009b).

Another reason for studying California is that, according to National Ambient Air Quality Standards established by the federal Clean Air Act, the state has an air quality problem. The EPA currently designates several California regions as nonattainment areas for particulate matter and ozone, and the population living in these nonattainment areas exceeds that of any other state by a large margin (EPA 2009b). In many of these areas the pollution conditions are identified by the EPA as "serious" or "severe." Most of the affected population live in the South Coast Air Basin (comprising parts of Los Angeles, Orange, Riverside, and San Bernardino counties), but there are several other highly populous nonattainment areas, including the San Joaquin Valley, the San Francisco Bay area, the Sacramento Metro area, and San Diego.

A vast body of epidemiological research has established the relationship between air pollution and human health.[2] There is a great deal of variety among studies. Geographical region(s), population(s), model functional form, model pollutants and covariates, and definitions of adverse health effects—among other dimensions—can and do vary.

Moolgavkar (2000b) is one example. This study analyzes the impact of particulate matter with an aerodynamic diameter of less than or equal to 2.5 micrometers (which we abbreviate as PM2.5) and other pollutant levels on daily hospital admissions for cardiovascular causes in metropolitan Los Angeles, Phoenix, and Chicago from 1987 to 1995. Its analyses account for time-varying confounding factors, such as temperature and relative humidity. Some analyses also account for multiple pollutants simultaneously. Such an approach can be useful in isolating the impact of a specific pollutant, because pollutant levels may be correlated. However, if pollutant levels are highly correlated, isolating the impacts of specific pollutants may not be feasible.

More generally, the presence of ambient PM2.5 and ozone lead to a raft of adverse health effects, referred to as "endpoints." PM2.5 has been shown to contribute significantly to such endpoints as adult and neonatal mortality, heart attacks, increased hospital and doctor visits, acute and chronic bronchitis, and asthma attacks (EPA 2006b). Ozone has been shown to contribute significantly to several respiratory endpoints, including respiratory-related hospital and ER visits, asthma attacks, minor restricted-activity days, and premature mortality from both acute and chronic exposure (Bell, Peng, et al. 2006; EPA 2008c; Jerrett, Burnett, et al. 2009). Concentrations of both pollutants vary by season. Ozone levels peak in the summer: Using daily 8-hour maximum concentration, the average July ozone level in the median California zip code is roughly twice that of January. The variation in PM2.5 is less dramatic for the typical Californian community, but the zip code at the 95th percentile of PM2.5 pollution experiences 57% higher levels of PM2.5 in January as compared with July.

While our study focuses on pollution-related hospital care, it is important to acknowledge that there are many other adverse health effects associated with these pollutants. Table 2.2 lists the health endpoints known to be associated with ozone and PM2.5, as identified by the EPA (2008c); other associated endpoints, whose relationships to the pollutants are uncertain or whose effects have not yet been well quantified (such as loss of lung function), are not listed.

---

[2] Holgate, Koren, et al. (1999) is a useful reference, in addition to those identified in the remainder of this chapter.

Table 2.2
Known and Quantified Health Endpoints Associated with PM2.5 and Ozone

PM2.5 health effects	Premature mortality
	Chronic and acute bronchitis
	Respiratory hospital admissions
	Cardiovascular hospital admissions
	ER visits for asthma
	Heart attacks (myocardial infarction)
	Lower and upper respiratory illness
	Minor restricted-activity days
	Work loss days
	Asthma exacerbations (asthmatic population)
	Respiratory symptoms (asthmatic population)
	Infant mortality
Ozone health effects	Premature mortality: short-term exposures
	Respiratory hospital admissions
	ER visits for asthma
	Minor restricted-activity days
	School loss days
	Asthma attacks
	Acute respiratory symptoms

NOTE: Shading indicates endpoints included in this study.

For the events we consider, spending—our focus here—does not capture significant social costs associated with these events. In particular, some of the hospital admissions that we study result in mortality, the true costs of which substantially exceed the amount that health care payers spend. Furthermore, air pollution results in medical spending outside of hospitals (for example, in doctors' offices); we were unable to examine who paid for this care. Hall, Brajer, et al. (2008a) quantifies the incidence of a broad range of pollution effects in two populous and polluted regions of California and includes estimates of social costs.[3]

Also beyond the scope of this study is an estimation of the cost to reduce pollution levels to national standards; EPA (2006b, 2008c) may be consulted on this issue.

To enforce air pollution standards, pollution data are collected at monitor sites throughout California and the country according to federal guidelines. Attainment status is determined using measures called "design values," which are derived from the raw data collected at monitors. When the design value for a pollutant exceeds the attainment threshold at a particular monitor, state bodies collaborate with the EPA to identify appropriate geographic boundaries for the nonattainment area around that monitor, and states must take a series of steps—

[3] This study also reviews methods for valuing these harms, including cost of illness, market-based, and contingent valuation approaches.

including obtaining approval from the EPA for a remedial "State Implementation Plan"—to ensure that air quality in the affected region is improved so as to meet the standard.

The most recent federal standard governing ozone is a 2008 revision that identifies the 8-hour ozone design value to be the 3-year average of the annual fourth-highest daily maximum 8-hour average ozone concentration and sets the attainment threshold at 75 parts per billion (EPA 2008c). Using the standard's prescribed rounding rules, a region's attainment is achieved when no monitor in that region reports a design value above 75.49 parts per billion.

The federal standard for particulate matter regulates two subclasses of the pollutant: particulate matter with an aerodynamic diameter of less than or equal to 2.5 micrometers, and particulate matter with an aerodynamic diameter of less than or equal to 10 micrometers, abbreviated as PM2.5 and PM10, respectively. Our analysis examines PM2.5, for which the federal standard is more stringent (Hall, Brajer, et al. 2008a, 2008b).

The most recent federal standard governing PM2.5 is the 2006 revision, which identifies the PM2.5 design value to be the 3-year average of the annual 98th percentile 24-hour average concentration and sets the attainment threshold at 35 micrograms per cubic meter. That is, attainment is achieved when no monitor in an area reports a design value above 35 micrograms per cubic meter (EPA 2009a). There is a second design value for PM2.5 based on average annual concentrations, which was not changed in the 2006 PM2.5 revision. Of the two, the newer 24-hour design value standard is more stringent, and we use it.[4]

The size and boundary characteristics of nonattainment areas recommended by the California Air Resource Board (and ultimately approved by the EPA) sometimes differ between PM2.5 and ozone. Ozone nonattainment areas follow the boundaries of "ozone 8-hour planning areas," which typically cover multiple counties or parts of counties. For example, the Sacramento Metro Area planning area encompasses Sacramento, Yolo, eastern Solano, southern Sutter, and western portions of El Dorado and Placer counties. Like ozone, PM2.5 nonattainment regions are determined on a case-by-case basis and use similar criteria, although the resulting nonattainment areas are sometimes different from those for ozone.

The full criteria for determining PM2.5 and ozone nonattainment boundaries are described in EPA (2008b) and EPA (2007), respectively. Both guidelines recommend a nine-point analysis, considering air quality data, emissions data, population density and degree of urbanization, traffic and commuting patterns, growth rates and patterns, meteorology, geography and topography, jurisdictional boundaries, and level of control of emission sources. As described in detail below, the choice of attainment boundaries is important to our results; our analysis uses the actual nonattainment regions determined by the EPA based on 2005–2007 pollution levels.

The state of California sets its own standards for PM2.5 and ozone (California Air Resources Board 2002a, 2002b, 2005a, 2005b). In particular, the state's daily maximum ozone concentration is 70 parts per billion, which is more stringent than the federal standard. California's 24-hour average PM2.5 standard is identical to the federal standard.[5] We consider these state standards in the next chapter.

[4] Under the linear rollback model described in the next chapter, the percentage reduction in PM2.5 levels is larger for the daily average standard than for the annual standard in all nonattainment areas.

[5] Under the linear rollback model described in the next chapter, the percentage reduction in PM2.5 levels is larger for California's daily average standard than for its annual standard in all nonattainment areas other than the South Coast Air Basin.

CHAPTER THREE

Analytic Approach

In order to assess the impact of poor air quality on health care purchasers, we considered a scenario in which clean air standards were met throughout California over 2005–2007. Given the available data, we analyzed the cost of not meeting air standards, and the resulting reductions in spending on hospital care if the standards had been met. The analysis involved three steps:

1. Measure the reductions in PM2.5 and ozone concentration levels throughout California for our air quality improvement scenario.
2. Determine the potential reductions in medical care delivered at specific hospitals.
3. Quantify potential spending reductions by type of payer/purchaser.

In the first step, we measured the reductions in ambient pollution levels needed within nonattainment areas for all of California to have met standards for PM2.5 and ozone based on 2005–2007 design values. Our main analysis used the federal standards described in the preceding chapter; we also considered California's clean-air standards. We then mapped the resulting pollution levels at monitor sites to pollution levels within zip codes; residential zip codes are included in data on hospital admissions and ER visits.

In the second step, we linked the decrease in ambient pollution levels within zip codes to potential reductions in treatment episodes—for hospital-related "endpoints" such as ozone-related asthma ER visits—throughout the state. We used concentration-response functions from epidemiological studies used in regulatory assessments of the benefits of air quality improvements to determine the number of health events requiring hospital care that would have been prevented within each zip code.[1]

In this step, we used information on the hospitals at which patients within each zip code received care in order to identify where the treatment episodes would have been prevented. This approach deals with the strong relationship between residential location and site of hospital care (Luft, Garnick, et al. 1990). Our hospital data include the calendar quarter of the care; we describe below how we reconcile the different frequencies of the pollution and hospital data. We also describe how we deal with confounding factors, such as socioeconomic status.

In the third and final step, we quantified the potential reductions in spending by type of payer. Our data on hospital admissions include charges based on full established rates. We determined actual spending using information for each hospital on gross and net inpatient revenue by type of payer.

[1] The epidemiological evidence (described below) does not show that air pollution becomes harmful only above thresholds equal to the regulatory standards, so it is likely that the number of events potentially prevented would be even larger if standards for air quality were exceeded (rather than just met).

In the next section, we describe the data sources used in the analysis. We then describe each analysis step in greater detail.

Data Sources

We used a variety of data sets in our analysis. The two most important concern pollution levels and hospital care.

Pollution data were obtained from the California Ambient Air Quality Data 1980–2007 DVD, released by the California Air Resources (California Air Resources Board 2009). This data set reports monitor locations (i.e., geo-coordinates); monitor-level concentrations of various pollutants at a daily and sometime hourly frequency; and design values for various relevant geographic regions (e.g., ozone 8-hour planning areas).

Information on hospital care was obtained from three data sets from the California Office of Statewide Health Planning and Development. First, we used the public versions of the Patient Discharge Data File (California Office of Statewide Health Planning and Development 2009c). For each hospital admission, these data report primary diagnosis, expected source of payment, charges at fully established rates, five-digit zip code of residence, quarter of care, patient age, and race/ethnicity. Second, we used public versions of the Emergency Department and Ambulatory Surgery Data File (California Office of Statewide Health Planning and Development 2009a). For each ER visit, these data report the fields just identified for the discharge data, with the exception of charges. ER visits that resulted in admissions are excluded from the ER data file but included in the discharge data. We used the three most recent years of the inpatient and ER data, 2005–2007. Third, we used Hospital Quarterly Financial and Utilization Data Files for the same period (California Office of Statewide Health Planning and Development 2009b). These data report gross and net inpatient revenue by payer type.

Zip code centroid locations were taken from a December 2007 geographic database (ZIPList5 2009). Finally, data on poverty status by zip code were obtained from the 2000 Census.[2]

Analysis Step 1: Measuring Reductions in Pollution Levels

In this initial step, we first measured the reductions in ambient pollution levels needed within nonattainment areas for all of California to have met federal or state standards for PM2.5 and ozone over 2005–2007.

To do so, we used a linear rollback model (EPA 1999a, 2003; Hall, Brajer, et al. 2008a, 2008b). Under this model, pollution levels in excess of the background level (i.e., absent human causes) are reduced by the same percentage.

Under this approach, pollution is reduced throughout a nonattainment area. Regional approaches to pollution control could generate such a pattern. At the same time, an equal percentage reduction throughout an area corresponds to a larger absolute reduction in "hot spots" with relatively high pollution levels. Targeted pollution control strategies could generate this

---

[2] The relevant Census geography is the Zip Code Tabulation Area, which is defined to approximate U.S. Postal Service zip codes.

pattern. Modeling rollback is complex and contingent on the pollution control strategy (Hall, Brajer, et al. 2008a). The strategy that would be used to meet clean air standards is unknown, and our approach should be viewed as an approximation, albeit a reasonable one.

Within each nonattainment area, the percentage reduction is chosen based on the pollution monitor with the maximum design value. Given the percentage reduction, this monitor's design value exactly equals the attainment threshold.[3]

The mathematical representation of the model is as follows:

$$L_{md}^{rollback} = \begin{cases} L^{background} + \dfrac{P}{100}\left(L_{md}^{actual} - L^{background}\right) & \text{if } L_{md}^{actual} \geq L^{background} \\ L_{md}^{actual} & \text{if } L_{md}^{actual} < L^{background} \end{cases} \tag{1}$$

where:

$L_{md}^{rollback}$ = rollback pollution level at monitor m on day d

L_{md}^{actual} = actual pollution level

$L^{background}$ = background level

P = percentage rollback, or $100\left(\dfrac{L^{\text{attain}} - L^{background}}{L^{\max} - L^{background}}\right)$

L^{attain} = attainment threshold

$L^{\max}$ = maximum design value in nonattainment area.

Pollution levels were defined as the daily average concentration level for PM2.5 and the 8-hour daily maximum for ozone, as in the National Ambient Air Quality Standards (see Chapter Two). It is therefore natural to consider pollution rollback at a daily frequency. Ozone levels were typically available at monitors every day; PM2.5 values were often measured only once every three days. We discuss how we deal with missing data in the next analysis section.

Background pollution levels were based on recommendations in recent EPA reports (EPA 2006a, 2008d). Levels included 1.01 micrograms per cubic meter for PM2.5 in northern California, 0.84 micrograms per cubic meter for PM2.5 in southern California, and 30 parts per billion for ozone throughout the state. The PM2.5 levels are based on annual mean values for the Northwest and southern California regions of the United States. The ozone level is the midpoint of a range for the western United States (Fiore, Jacob, et al. 2002). Fiore, Jacob, et al. (2002) observe that background ozone generally drops below 15 parts per billion where actual ozone levels are likely to be high. Thus, the use of 30 parts per billion is conservative with respect to the magnitude of ozone reductions resulting from the model.

For nonattainment areas, we used the most recent designations made by the California Air Resources Board and the EPA under the 2008 revision of the ozone standard and the 2006

---

[3] Design values and attainment levels were reviewed in Chapter Two.

revision of the PM2.5 standard (EPA 2008a, 2008e; California Air Resources Board 2009). Chapter Two described California nonattainment areas.

We mapped the resulting pollution concentration levels at monitor sites to daily pollution levels within zip codes. For each zip code, we first identified monitors located within 20 miles and with valid data on a particular day. (Zip codes with no monitors of a pollutant within this distance throughout 2005–2007 were excluded from the analysis of that pollutant.[4]) We then averaged the monitor-level pollution levels, with inverse distances as weights (Neidell 2004). Distance was measured by applying the great circle formula to the geo-coordinates of a zip code's centroid and those of monitors.[5]

Analysis Step 2: Determining Reductions in Hospital Care

In this analysis step, we linked the decrease in pollution levels to reductions in care at hospitals throughout the state. We first linked decreased pollution levels to reductions in health events requiring hospital care—that is, to reductions in "endpoint" events—within residential zip codes. Our focus on exposure close to home is similar to others (Hall, Brajer, et al. 2008a, 2008b); comprehensive information on exposure to pollution at a substantial distance from home is unavailable in our setting. We then linked endpoint reductions within zip codes to reductions in care at specific hospitals. In both cases, we used the hospital care data described earlier in this chapter.

To link decreased pollution levels to endpoint reductions within zip codes, we also used epidemiological studies on the health effects of PM2.5 and ozone. Such studies quantify the parameters of concentration-response functions. Table 3.1 identifies the studies used in our main analyses and describes their endpoints and populations (e.g., pneumonia admissions for persons 65 and older).

The EPA has reviewed the epidemiological literatures on the health effects of PM2.5 and ozone (EPA 2006b, 2008c). The agency prefers peer-reviewed, U.S.-based, multicity studies that cover the broadest potentially exposed and sensitive populations and that use multivariate models to account for covariance between pollutants as well as other factors (EPA 2003, 2004).[6] In addition, the EPA values recent evidence, but also longer timeframes (and larger samples) that provide greater statistical precision. Based on these criteria, studies have been selected for the agency's Environmental Benefits Mapping and Analysis Program, or BenMAP (Abt Associates 2008).

In order for our findings to be as credible as possible, we used studies that have been deemed reliable for regulatory analyses of the benefits of improved air quality. In particular, all of the studies in Table 3.1 are used in the current version of BenMAP, with the exception of

[4] For ozone, 2.2% of California's population was thereby excluded, based on 2000 Census data. For PM2.5, 8.1% was excluded. In both cases, excluded zip codes tended to be lightly populated.

[5] While California's air pollution monitor network is extensive, some sources of pollution may be located significant distances from a monitor. In these cases, the impact of pollution on communities near the unmonitored "hot spot" will likely be underestimated by our spatial averaging approach.

[6] In addition, the EPA prefers studies that have large, and thus informative, samples.

Table 3.1
Concentration-Response Beta Values from the Epidemiological Literature

Study	Pollutant	Endpoint	ICD-9 codes	Population	Metric	Beta[a]
Thurston and Ito 1999[b]	Ozone	Acute bronchitis, pneumonia, or COPD	466, 480–486, 490–496	All ages	1-hr daily max	0.001655
Peel, Tolbert, et al. 2005; Wilson, Wake, et al. 2005	Ozone	Asthma ER visit	493	All ages	8-hr daily max	0.001215
Sheppard 2003	PM2.5	Asthma admission	493	Age 0–64	24-hr avg	0.003324
Moolgavkar 2000a	PM2.5	COPD excl. asthma admission[c]	490–492, 494–496	Age 18–64[d]	24-hr avg	0.0022
Ito 2003; Moolgavkar 2003	PM2.5	COPD admission	490–496	65 and older	24-hr avg	0.001809
Ito 2003	PM2.5	Pneumonia admission	480–486	65 and older	24-hr avg	0.00397
Moolgavkar 2000b	PM2.5	All cardiovascular admissions	390–429	Age 18–64[e]	24-hr avg	0.0014
Moolgavkar 2003	PM2.5	All cardiovascular admissions	390–429	65 and older	24-hr avg	0.00158
Norris, YoungPong, et al. 1999	PM2.5	Asthma ER visit	493	17 and younger	24-hr avg	0.016527

NOTES: ICD-9 = International Statistical Classification of Diseases and Related Health Problems. COPD = chronic obstructive pulmonary disease. The National Heart, Lung, and Blood Institute defines COPD as a "serious lung disease which makes it hard to breath. Also known . . . as emphysema or chronic bronchitis, COPD is now the 4th leading cause of death in the U.S." (National Heart, Lung, and Blood Institute, undated).

[a] All concentration-response functions in our analysis are log-linear (exponential) in form, as described in detail in our description of analysis step 2.

[b] This study was the only one used in our analysis that was not also identified in EPA (2008c). However, the authors are established researchers in this area, the study was published in a highly regarded text, and it has been used in other recent benefits assessments: Hall, Brajer, et al. (2008a, 2008b).

[c] Moolgavkar (2000a) includes asthma admissions (ICD 493) in his analysis. We exclude it, following (Abt Associates 2008), in order to prevent double-counting of nonelderly asthma admissions, for which a separate study, (Sheppard 2003), is used.

[d] Moolgavkar (2000a) uses an age category of 20–64 in his analysis. We modify this to 18–64, following (Abt Associates 2008), in order to better match to Office of Statewide Health Planning and Development age categories.

[e] Here we again modify Moolgavkar's (2005b) original age range from 20–64 to 18–64.

Thurston and Ito (1999). This latter study has been used by the California Air Resources Board and others (California Air Resources Board 2006; Hall, Brajer, et al. 2008a, 2008b).[7]

This approach has potential limitations. First, we did not use recent evidence that has yet to be incorporated into BenMAP, but may be in the future. Second, some endpoints include only the elderly or nonelderly, because we lacked studies specific to both groups.[8] As an example, PM2.5 may lead to hospital admissions among 64-year-olds, as well as those 65 and older. We were unwilling to extrapolate beyond the populations studied (in this case, age 65 and older), out of a concern that our findings would be less reliable and credible. The impact of

[7] We did not use all BenMAP studies, in particular, Jaffe, Singer, et al. (2003). This study of the impact of ozone on asthma ER visits analyzes an age range (5–34) that is spanned by the age ranges in Peel, Tolbert et al. (2005) and Wilson, Wake, et al. (2005). These latter studies are also more recent than Jaffe, Singer, et al. (2003).

[8] For other endpoints, such as cardiovascular admissions, studies considered all age groups.

improved air quality on hospital spending may therefore be understated, particularly for less vulnerable populations (in particular, nonelderly adults).

For some of the endpoints analyzed, there were multiple studies that used distinct yet overlapping sets of principal diagnoses (i.e., ICD-9 codes).[9] Combining the results of such studies would have resulted in some double counting. In our main analyses, we used the more inclusive set of diagnoses. We assessed this approach with a sensitivity analysis described below.

For all of the selected studies, the concentration-response function is exponential. Mathematically, the number of endpoints for the population of a zip code on any day can thus be represented as[10]:

$$E_{zd} = \gamma_z \cdot \exp(\beta \tilde{P}_{zd}) \cdot \varepsilon_{zd} \quad (2)$$

where:

E_{zd}	= actual number of endpoint events in zip code *z* on day *d*
β	= parameter that quantifies the percentage impact of 1-unit change in pollution on the expected number of endpoint events
$\tilde{P}_{zd}$	= pollution level in zip code *z* on day *d*, net of mean level
ε_{zd}	= random component of endpoint events, with mean 1
γ_z	= expected number of endpoint events on a day with average pollution, given mean pollution (i.e., $\tilde{P}_{zd} = 0$).

This (and subsequent) equations were applied separately to each of the endpoints in Table 3.1.

The expected change in endpoint events given a change in pollution dP_{zd} is then:

$$dE_{zd} = \beta\gamma_z \cdot \exp(\beta \tilde{P}_{zd}) \cdot dP_{zd} \quad (3)$$

In this equation, dP_{zd} was calculated as in analysis step 1. Where the epidemiological study metric differed from the metric of the attainment threshold, we applied the linear rollback to the epidemiological metric for the sake of consistency.[11] $\tilde{P}_{zd}$ was calculated as a spatial average of actual pollution monitor data (see preceding section on prior analysis step).

---

[9] For example, Moolgavkar (2003) studied the relationship between PM2.5 and all cardiovascular admissions, using ICD-9 codes 390–429, but Ito (2003)—while also reporting an intent to study the relationship between PM2.5 and cardiovascular admissions—looked at only ICD-9 codes 410–414 (ischemic heart disease), 427 (dysrhythmia), and 428 (heart failure). The disagreement in this case is not as large as it may seem; Ito's small subset of cardiovascular codes accounted for approximately 80% of the admissions matching Moolgavkar's codes.

[10] The relationship in Equation 1 is typically (and equivalently) stated as

$$\log E_{zd} = \log \gamma_z + \beta \tilde{P}_{zd} + \log \varepsilon_{zd}.$$

Our specification allows for a convenient interpretation of γ_z.

[11] For example, Table 3.1 shows that Thurston and Ito (1999) used a 1-hour daily maximum for ozone.

Table 3.1 reports the β parameter values used.[12] As a matter interpretation, the impact of a one-unit increase in the pollution level on the number of endpoints is equal to β multiplied by 100; the impact of a one-unit decrease is just the negative of this value. For example, a 1 microgram per cubic meter decrease in daily average PM2.5 leads to a 0.3324% decrease in the number of hospital admissions for asthma (Sheppard 2003).

Where multiple epidemiological studies used the same diagnoses for an endpoint, we took an inverse-variance-weighted average of their β estimates, as described in EPA (2005). Where multiple studies used disjointed sets of diagnoses for an endpoint, our analysis applied the β estimate for each to its diagnoses and summed across studies.

We aggregated Equation 2 up to a quarterly frequency, because hospital care data report the quarter of care:

$$dE_z = \frac{1}{\lambda_z} \sum_d \beta \gamma_z \cdot \exp(\beta \bar{P}_{zd}) \cdot dP_{zd} \tag{4}$$

where

λ_z = the fraction of days in quarter for which pollution data were available, i.e., $N^{data} / N^{quarter}$

N^{data} = number of days of with pollution data

$N^{quarter}$ = number of days in quarter.

The summation in Equation 3 can be taken only over days for which pollution data were available. The λ_z term corrects for missing data. This approach assumes that missing data are unrelated to pollution levels.

The expected number of endpoint events γ_z in a particular quarter is unknown.[13] We estimated it as:

$$\hat{\gamma}_z = \frac{\lambda_z E_z}{\sum_d \exp(\beta \bar{P}_{zd})} \tag{5}$$

where E_z is the actual number of endpoint events in a zip code in the hospital care data. The issue of missing hospital care data is discussed in Appendix C.

This approach to unobserved determinants of endpoint events is useful. In general, many factors, including socioeconomic status, affect health (see, e.g., Smith [1999]). Equation 4 allows these factors to vary arbitrarily across zip codes, mitigating the common and often serious concern about confounding bias. Equation 4 also allows for arbitrary variation across the 12 quarters analyzed, and hence for seasonal patterns in the expected number of events on an average pollution day. For example, pneumonia admissions are more likely in winter.

---

[12] Abt Associates (2008) transformed the results of the EPA's preferred studies into standard β parameters.

[13] The mean daily pollution level that defines $\bar{P}_{zd}$ is likewise specific to the quarter.

The resulting zip code–level endpoint reductions did not distinguish among payer types. To make this important distinction, we used patterns of care with respect to payers in the hospital care data. In each zip code, we assumed that the number of endpoints for a payer type decreased in proportion to the zip code's share of admissions or ER visits with that payer type.

To illustrate, consider a hypothetical example. Suppose that in a particular quarter of a particular year, 100 nonelderly (i.e., under age 65) residents of a zip code suffered episodes of asthma serious enough to be admitted to a hospital. Suppose that 80% of these patients used a private third-party insurer as their primary payer, while the remainder used Medicaid. Finally, suppose that our pollution reduction analysis showed that five asthma admissions originating in the zip code would have been prevented by reductions in PM2.5 levels, had the federal standard been met. Based on the actual distribution of nonelderly asthma admissions, we would have estimated that private third-party insurers would have paid for four of the admissions (80% of five) and that Medicaid would have paid for the remaining admission.

This approach is consistent with the uniform impacts of pollution on endpoints in the epidemiological studies used (that is, these studies did not estimate distinct β parameters for different populations).[14] Zip code shares were specific not only to the endpoint, but also to the population and calendar quarter. Appendix D describes the payer categories in the discharge and ER visit data.

To link endpoint reductions to reductions in care at specific hospitals, we used the actual geographic patterns of care in the hospital care data. In particular, we assumed that the number of endpoints at a hospital originating from a zip code decreased in proportion to the hospital's share of admissions or ER visits from that zip code. We used hospital shares that were specific not only to the endpoint, but also to the population, detailed payer type, and calendar quarter. Thus, reductions in care were specific to type of payer. This approach requires that the site of hospital care be unrelated to pollution levels, given an endpoint, population, payer type, and calendar quarter.

Analysis Step 3: Quantifying Spending Reductions by Payer Type

In the prior analysis step, we determined the reductions in care delivered at each hospital under our air quality improvement scenario. In this step, we quantified the resulting reductions in spending on hospital care by type of payer.

The discharge data include charges based on each hospital's full established rates, but not actual spending. We used the reduction in charges to determine the reduction in actual spending by payer type at hospitals, for each pollutant, endpoint, and population combination. To determine reductions in charges, we summed the charges for all admissions matching the endpoint, population, and payer type. We multiplied this total by the proportion of events (defined by endpoint and population) that would have been prevented at each hospital due to the reduction in PM2.5 or ozone pollution levels. Our approach requires that average charges

[14] There is some evidence that the health impact of pollution may be larger or smaller according to socioeconomic status (O'Neill, Jerrett, et al. 2003). None of the studies that met our selection criteria allowed β to vary with patient characteristics. Thus, the percentage impact of pollution cannot vary with such characteristics. However, the *absolute* impact can vary with average patient characteristics within zip codes through the γ_z term in Equation 4.

be similar for potentially prevented events and other admissions for the same endpoint, populations, and payer type.

To determine potential reductions in spending, we measured the discount off charges for actual payments at hospitals. This discount is equal to gross inpatient revenue less net inpatient revenue, divided by gross revenue, all by payer type. We determined reductions in actual spending by discounting total reductions in charges. Revenue information is reported on a quarterly basis; we therefore calculated and applied discounts at a quarterly frequency. Our approach requires that discount rates at a hospital be similar across admissions with the same payer type.

Out of 416 California hospitals with any of the endpoints analyzed during the study period, 35 did not report charges for any inpatient admissions.[15] Spending reductions cannot be determined directly by discounting charges for these hospitals.[16] We estimated spending for the associated admissions based on average spending among patients with the same diagnoses, zip code, payer type, and calendar year.

More than 90% of the admissions to hospitals without charges were for Kaiser Permanente, an integrated delivery system that both delivers and finances medical care. Avoided spending serves as an approximation to avoided costs at these hospitals.

The ER visit dataset does not include charges. For asthma-related ER visits (the only ER endpoint examined), we measured actual spending based on the EPA's Cost of Illness Handbook (EPA 1999b). The handbook reports an average amount paid by Medicare under its prospective payment system of $442.84 in 1999. We accounted for increases in medical costs between 1999 and the study period by inflating this spending level using the Consumer Price Index for Medical Care.[17]

Finally, Appendix C describes the payer category assignments made between the discharge and ER visit data and the quarterly hospital financial reports.

Sensitivity Analyses

To test the sensitivity of our results to parametric uncertainty and different methodological decisions, we examined two aspects of our analysis more closely: (1) uncertainty about the "true" relationship between ambient pollutant levels and health endpoints, as represented by the concentration-response function β values; and (2) the decision to use studies with more inclusive endpoint and population definitions as the basis for our pollutant concentration-response functions.

Our first sensitivity analysis described the breadth of results associated with uncertainty about our C-R function β estimates. For each pollutant, endpoint, and affected population,

[15] Charges are reported for virtually all other admissions at all other hospitals.

[16] Our analysis can and does include the number of admissions to the 35 hospitals that would have been prevented by improved air quality.

[17] For comprehensiveness, we estimated charges for ER visits by "reverse discounting" spending using the 2005–2007 volume-weighted average discount enjoyed by Medicare for fee-for-service patients (77.14%). Annual values of spending and charges for asthma ER visits are available from the authors upon request.

we determined low and high estimates of event and cost savings based on the lower and upper 95% confidence interval boundary value of the β derived from the epidemiological literature.[18]

To calculate the variance of the aggregate results, we first estimate the variance of each row's results by finding the difference between the upper and lower 95% confidence interval bounds and using a normal approximation (i.e., we assume that the distance between the bounds is $1.96 \times 2 = 3.92$ standard deviations). We assume the β estimates are independent (and so assume that each row's results are independent); in that case, the variance of the sum of the results across all rows is equal to the sum of the rows' variances. The upper and lower 95% confidence interval bounds of the aggregate results are then calculated by applying a normal approximation to this aggregate estimated variance.

Our second sensitivity analysis concerned endpoint and population definitions. As noted earlier in this chapter, we used the more inclusive epidemiological studies as the basis for our central savings estimates. The alternative was to use narrower studies that focused on smaller ranges of ICD-9 codes and/or affected age groups and aggregate those results.

Table 3.2 reports the epidemiological studies used. (Note that some studies were used in both approaches.) Consider cardiovascular admissions among the elderly. The sensitivity analysis included ischemic heart disease, dysrhythmia, and heart failure, based on ICD-9 codes of 411–414, 429, and 428, respectively. The benchmark analysis used a broader range of codes (390–429) that also included, for example, hypertensive disease.

---

[18] In other words, the upper and lower bound estimates for our results are calculated by redoing our analysis using the 95% confidence interval upper and lower bound estimates of each β. Note that, because the concentration-response functions we use are convex, the upper and lower bounds on the results we calculate will not be symmetrical around the central result estimate; however, the probability mass contained within the upper and lower bounds of the β estimate is maintained in the bounds for our results, so it is reasonable to also think of these as 95% confidence interval bounds.

Table 3.2
Epidemiological Studies Used for Sensitivity Analysis Based on Narrower ICD-9 Code and/or Age Ranges

Contributing Study	Pollutant	Endpoint	ICD-9 codes	Population	Metric	Beta[a]
Moolgavkar, Luebeck, et al. 1997; Ito 2003; Schwartz 1994	Ozone	Pneumonia admission	480–486	65 and older	24-hr avg	0.004313463
Schwartz 1994	Ozone	COPD admission, excluding asthma	490–492, 494–496	65 and older	24-hr avg	0.005523
Burnett, Smith-Doiron, et al. 2001[b]	Ozone	Croup, acute bronchitis, pneumonia, or asthma admission	464.4, 466, 480–486, 493	Age 0–1	1-hr max (5-day moving average)	0.0073
Peel, Tolbert, et al. 2005; Wilson, Wake, et al. 2005	Ozone	Asthma ER visit	493	All ages	8-hr daily max	0.001215
Ito 2003; Moolgavkar 2003	PM2.5	COPD admission	490–496	65 and older	24-hr avg	0.001809
Moolgavkar 2000a	PM2.5	COPD excl. asthma admission[c]	490–492, 494–496	Age 18–64[d]	24-hr avg	0.0022
Ito 2003	PM2.5	Pneumonia admission	480–486	65 and older	24-hr avg	0.00397
Sheppard 2003	PM2.5	Asthma admission	493	64 and younger	24-hr avg	0.003324
Norris, YoungPong, et al. 1999	PM2.5	Asthma ER visit	493	17 and younger	24-hr avg	0.016527
Ito 2003	PM2.5	Ischemic heart disease	411–414	65 and older	24-hr avg	0.001435
Ito 2003	PM2.5	Dysrhythmia	429	65 and older	24-hr avg	0.001249
Ito 2003	PM2.5	Heart failure	428	65 and older	24-hr avg	0.003074
Moolgavkar 2000b	PM2.5	All cardiovascular admissions	390–429	Age 18–64[e]	24-hr avg	0.0014

[a] All concentration-response functions in our analysis are log-linear (exponential) in form.

[b] This relationship was found only between May and August; Burnett finds no relationship between ozone and infant respiratory admissions in colder months, when ozone levels are substantially lower.

[c] Moolgavkar (2000a) includes asthma admissions (ICD 493) in his analysis. We exclude it, following Abt Associates (2008), in order to prevent double-counting of nonelderly asthma admissions, for which a separate study, Sheppard (2003), is used.

[d] Moolgavkar (2000a) uses an age category of 20–64 in his analysis. We modify this to 18–64, following Abt Associates (2008), in order to better match to Office of Statewide Health Planning and Development age categories.

[e] Here we again modify Moolgavkar's (2000b) original age range from 20–64 to 18–64.

CHAPTER FOUR

Results

This chapter first describes our overall results, then reviews the sensitivity analyses, and finally presents several case studies of specific hospitals.

Overall Results

Table 4.1 summarizes our results by pollutant, health endpoint, and population. Results within regions and by detailed payer type appear in Appendix A, while results for some specific health plans appear in Appendix B.

Failing to meet federal clean air standards for PM2.5 and ozone caused an estimated 29,808 hospital admissions and ER visits throughout California over 2005–2007.[1] (To prevent double counting, hospital admissions were defined to include hospital encounters that began in the ER but that led to an admission.) PM2.5 accounts for 72.9% of these events. Hospital admissions account for 52.3%. People age 65 or older account for 13,083 events, while minors

Table 4.1
Air Pollution–Related Hospital Events, Spending, and Hospital Charges in California over 2005–2007 Caused by Failure to Meet Federal PM2.5 and Ozone Standards, by Pollutant, Endpoint, and Population

Pollutant	Endpoint	Population	Events	Spending	Hospital Charges
Ozone	Acute bronchitis, pneumonia, or COPD admission	All ages	6,056	$56,500,000	$226,000,000
PM2.5	Pneumonia admission	65 and older	2,517	$27,700,000	$123,000,000
PM2.5	COPD admission	65 and older	652	$5,634,450	$24,800,000
PM2.5	COPD admission excl. asthma	Age 18–64	306	$2,721,382	$10,900,000
PM2.5	Asthma admission	64 and younger	940	$5,575,469	$20,100,000
PM2.5	Any cardiovascular admission	65 and older	3,256	$47,700,000	$205,000,000
PM2.5	Any cardiovascular admission	Age 18–64	1,864	$35,100,000	$120,000,000
Ozone	Asthma ER visit	All ages	2,027	$1,768,883	$5,271,011
PM2.5	Asthma ER visit	17 and younger	12,190	$10,400,000	$31,700,000
Total			**29,808**	**$193,100,184**	**$766,771,011**

[1] The potential reductions in events for the South Coast and San Joaquin Valley air basins estimated in our study are comparable to analogous estimates for those basins published in Hall, Brajer, et al. (2008a).

account for 14,373 events.[2] As we noted in the last chapter, our approach is conservative and likely to understate the number of events, particularly for persons age 18–64.

Overall spending on hospital care would have been $193 million lower if California had met federal clean air standards in this period.[3] To give some sense of this magnitude, the annual savings would be sufficient to pay for pediatric influenza vaccinations for 85% of California's under-15 population (Centers for Disease Control 2009; U.S. Census Bureau 2009). If state (rather than federal) clean air standards had been met, hospital spending would have been $204 million lower. The remainder of this chapter focuses on the scenario in which federal standards would have been met.

PM2.5 accounts for 69.8% of the overall potential spending reduction. Hospital admissions account for 93.7%. The hospital share of potential spending reductions is larger than the hospital share of events prevented, because hospital admissions are costlier than ER visits. Potential spending reductions on persons 65 and older account for 60.1% of the age-specific total. Overall, the potential reduction in spending reflects a 74.8% discount off hospital charges, based on fully established rates.[4]

The results by broad payer type are reported in Table 4.2. (Results for detailed payer types appear in Appendix A.) Public payers account for 62.3% of potentially prevented events. These events are split almost evenly between Medicare and Medi-Cal, as California's Medicaid program is known (31.0% and 30.1%, respectively). County indigent programs accounting for the rest (1.1%). The large Medicare share is attributable to the large share of potentially prevented events for persons 65 and older. Private third-party payers (including both managed-care and fee-for-service insurance plans) account for 30.3% of events. All other payer types—including "self-pay," that is, cases in which the patient is responsible for the majority of hospital charges incurred—account for the rest (7.4%). County indigent programs and self-pay include patients who lack health insurance.

Table 4.2
Potential Reductions in Events, Spending, and Hospital Charges in California over 2005–2007 Had Federal PM2.5 and Ozone Standards Been Met, by Payer Type

Payer	Reduction in Events	% of Total Event Reduction	Reduction in Spending	% of Total Spending Reduction	Reduction in Hospital Charges
Medicare	9,247	31.02	$103,600,000	53.60	$463,000,000
Medi-Cal	8,982	30.13	$27,292,199	14.14	$126,000,000
County indigent	335	1.12	$1,071,967	0.55	$7,612,133
Total public	18,564	62.28	$131,964,166	68.29	$596,612,133
Total private third-party	9,029	30.29	$55,879,780	28.90	$149,954,889
Total all other	2,216	7.43	$5,443,008	2.82	$20,919,389

NOTE: Medi-Cal is the name for California's Medicaid program.

[2] These numbers include the elderly and young within the "all ages" populations listed in Table 4.1. The patient age information in the hospital care data is used for this purpose.

[3] Ideally, these potential spending reductions could be compared with those in Fuchs and Frank (2002). The results are not comparable, because the pollution metrics differ.

[4] The overall discount is ($766,771,011 – $193,100,184)/$766,771,011, or 74.8%.

Public payers account for an even larger share of the overall potential reduction in spending on hospital care (68.3% versus 62.3%). The Medi-Cal share of potentially reduced spending is smaller than its share of prevented events (14.1% versus 30.1%). Yet the Medicare share of potentially reduced spending is much larger (53.6% versus 31.0%), because persons 65 and older are relatively likely to be admitted to the hospital (rather than treated in the ER) for pollution-related causes, and these admissions are costly.

Altogether, public spending on hospital care in California would have been $131,964,166 lower over 2005–2007 if federal clean air standards had been met. Private third-party payers account for 28.9% of the overall potential spending reduction, or $55,879,780.

Tables 4.1 and 4.2 include potential spending reductions for the 35 hospitals that did not report charges for hospital admissions.[5] As described in the preceding chapter, we used a different approach to estimate potential spending reductions for admissions to these hospitals. Overall, spending on admissions to these hospitals would have decreased by $27,700,000 if federal clean air standards had been met.

Sensitivity Analyses

We first assessed the sensitivity of our results to uncertainty regarding the magnitude of the impact of reduced air pollution on our study endpoints. The results are reported in Table 4.3. With 95% confidence, anywhere from 21,892 to 37,725 events requiring hospital care would

Table 4.3
95% Confidence Intervals for Potential Reductions in Events, Spending, and Hospital Charges in California over 2005–2007 Had Federal PM2.5 and Ozone Standards Been Met

Pollutant	Endpoint	Population	Reduction in Events	Reduction in Spending	Reduction in Hospital Charges
Ozone	Acute bronchitis, pneumonia, or COPD	All ages	3,554–8,583	$33,200,000–$80,100,000	$133,000,000–$320,000,000
PM2.5	Pneumonia admission	65 and older	445–4,683	$4,912,799–$51,600,000	$21,800,000–$229,000,000
PM2.5	COPD admission	65 and older	291–1,017	$2,516,082–$8,790,103	$11,100,000–$38,600,000
PM2.5	COPD admission excl. asthma	Age 18–64	105–512	$934,797–$4,542,988	$3,740,050–$18,200,000
PM2.5	Asthma admission	64 and younger	356–1,541	$2,111,007–$9,138,075	$7,594,736–$32,900,000
PM2.5	Any cardiovascular admission	65 and older	1,876–4,648	$27,500,000–$68,100,000	$118,000,000–$293,000,000
PM2.5	Any cardiovascular admission	Age 18–64	970–2,765	$18,300,000–$52,100,000	$62,500,000–$178,000,000
Ozone	Asthma ER visit	All ages	534–3,537	$465,893–$3,086,577	$1,388,160–$9,198,401
PM2.5	Asthma ER visit	17 and younger	5,825–19,442	$4,990,326–$16,600,000	$15,200,000–$50,600,000
All events			**21,892–37,725**	**$150,049,286–$236,151,082**	**$590,145,739–$943,396,283**

[5] Prevented events are included for these hospitals.

have been prevented if federal clean air standards had been met throughout California over 2005–2007. The 95% confidence interval on reduced spending ranged from $150 million to $236 million.

We then assessed the sensitivity of our results to alternative clinical definitions of the endpoints, as explained in the preceding chapter. These results are reported in Table 4.4. Based on these alternative definitions, 26,381 events would have been prevented, and hospital spending would have been reduced by $153 million.

Case Studies

We also determined the impact of improved air quality at specific hospitals. Five hospitals are presented here as case studies: Riverside Community Hospital, St. Agnes Medical Center, St. Francis Medical Center, Stanford University Hospital, and University of California–Davis Medical Center.

These case studies are a diverse group. We reviewed and qualitatively selected hospitals according to the following criteria: the scale of potential prevented events and spending reductions; geographic region; and payer and patient mix.

Figure 4.1 shows the number of potentially prevented events by patient zip code. These events are concentrated in the San Joaquin Valley and South Coast air basins. St. Agnes is located in the former, while Riverside Community and St. Francis are located in the latter. PM2.5 and ozone levels in these areas substantially exceed federal standards. A sizable number

Table 4.4
Potential Reductions in Events, Spending, and Hospital Charges in California over 2005–2007 Had Federal PM2.5 and Ozone Standards Been Met, Using Narrower ICD-9 Code and/or Age Ranges

Pollutant	Endpoint	Population	Reduction in Events	Reduction in Spending	Reduction in Hospital Charges
Ozone	Pneumonia admission	65 and older	862	$8,724,464	$37,700,000
Ozone	COPD admission excl. asthma	65 and older	514	$4,170,069	$17,800,000
Ozone	Croup, acute bronchitis, pneumonia, or asthma admission	1 and younger	1,067	$6,422,965	$21,300,000
PM2.5	Pneumonia admission	65 and older	2,517	$27,700,000	$123,000,000
PM2.5	COPD admission	65 and older	652	$5,634,450	$24,800,000
PM2.5	COPD admission excl. asthma	18–64	307	$2,721,382	$10,900,000
PM2.5	Asthma admission	64 and younger	940	$5,575,469	$20,100,000
PM2.5	Ischemic heart disease	65 and older	1,081	$20,200,000	$86,000,000
PM2.5	Dysrhythmia	65 and older	451	$4,497,242	$19,400,000
PM2.5	Heart failure	65 and older	1,909	$20,700,000	$91,000,000
PM2.5	Any cardiovascular admission	18–64	1,864	$35,100,000	$120,000,000
Ozone	Asthma ER visit	All ages	2,027	$1,768,883	$5,271,011
PM2.5	Asthma ER visit	17 and younger	12,190	$10,400,000	$31,700,000
Total			**26,381**	**$153,614,924**	**$608,971,011**

Figure 4.1
Potentially Prevented Events throughout California over 2005–2007 Had Federal PM2.5 and Ozone Standards Been Met, by Patient Zip Code

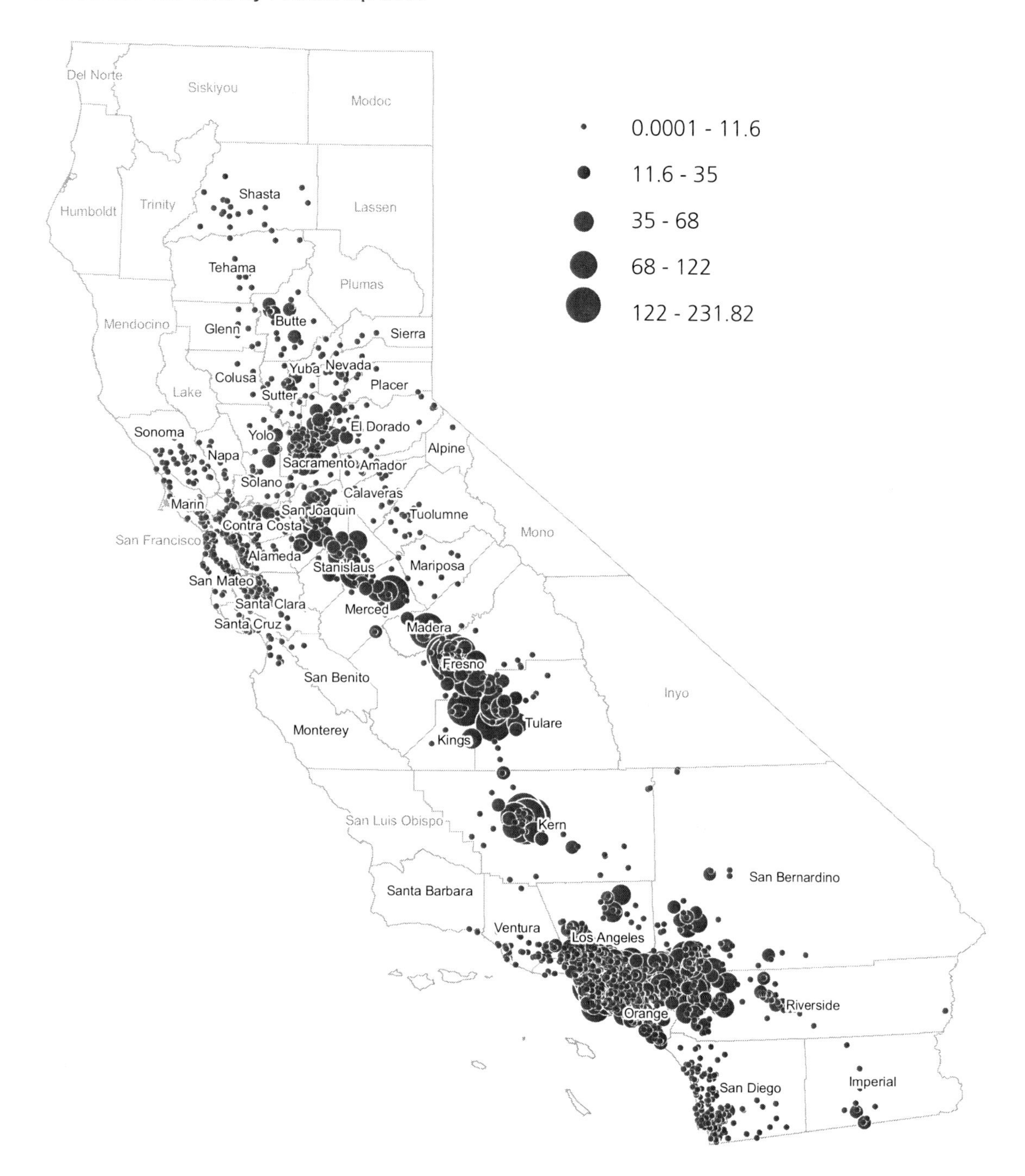

of events are also prevented in and near Sacramento, where the UC Davis Medical Center is located.

Stanford University Hospital is located in the San Francisco metropolitan area. Moreover, as Table 4.5 shows, private insurers were expected to pay most of the bill for 46% of patients at Stanford University Hospital, versus 31% for California as a whole. At the other extreme, private payers paid for only 14% of patients at St. Francis. Medi-Cal paid for 59% of patients,

Table 4.5
Characteristics of Case Study Hospitals, 2005–2007

Hospital	Riverside Community Hospital	St. Agnes Medical Center	St. Francis Medical Center	Stanford University Hospital	UC Davis Medical Center	All California Hospitals
Summary information						
City	Riverside	Fresno	Lynwood	Stanford	Sacramento	—
County	Riverside	Fresno	Los Angeles	Santa Clara	Sacramento	—
Annual discharges	18,903	24,396	22,841	22,788	29,282	7,248
Staffed beds	345	406	384	454	550	175
Teaching hospital	No	No	No	Yes	Yes	—
Discharges, by payer (%)						
Private third-party	37	30	14	46	35	31
Medicare	36	50	21	38	24	37
Medi-Cal	22	18	59	9	29	22
Other	5	2	7	7	13	10
Patient race/ethnicity (%)						
White	51	72	2	78	50	62
Black	7	4	20	5	12	7
Hispanic	38	21	77	7	18	24
Asian or Pacific Islander	1	3	0	10	5	5
American Indian	0	0	0	0	0	0
Other	3	0	1	0	15	2
Patient economic status, by income as percentage of Federal Poverty Level						
0–100% FPL	15	20	27	9	16	15
> 100% FPL	85	80	73	91	84	85

NOTES: Medi-Cal is the name for California's Medicaid program. See Table 4.4 for detailed payer types. Racial groupings include non-Hispanic persons of single race.

compared with a state average of 22%. Among the case study hospitals, the Medicare share was highest at St. Agnes (50%) and lowest at St. Francis (21%).

The racial composition of patients varied substantially across hospitals. Slightly more than three-quarters of patients were white at Stanford University Hospital, compared with 2% at St. Francis. African Americans were 20% of the patient population at St. Francis, compared with a statewide average of 7%. The proportion of Hispanics patients was well above average at St. Francis (77%) and at Riverside Community Hospital (38%).

The economic status of patients also varied widely. Statewide, 15% of patients have incomes below the federal poverty level. But at St. Francis, more than one-quarter of patients were poor; at Stanford University Hospital, fewer than 10% of patients were poor.

Figures 4.2 through 4.11 show the number of air pollution–related events at each of the five case-study hospitals:

At **Riverside Community Hospital**, 329 hospital admissions and ER visits would have been prevented had federal standards for PM2.5 and ozone been met during 2005–2007 (Figure 4.2). Private health insurers would pay most of the bill for almost half (149) of these patients. Medicare would be the next most frequent payer for these preventable events. Overall, spending would have been reduced by $2,015,880 (Figure 4.3). Medicare would have enjoyed the largest reduction (about $1,140,060), as these patients were relatively likely to have costly hospital stays, rather than ER visits. Private insurers would have saved about $708,700.

At **St. Agnes Medical Center** in Fresno, meeting federal air standards would have had even greater effects: 384 events would have been prevented (Figure 4.4) and $2,976,936 saved (Figure 4.5). More than half of the potentially prevented events (208) would have been paid for primarily by Medicare, consistent with its above-average importance at this hospital. Medicare would also have experienced the largest spending reduction ($1,913,116).

At **St. Francis Medical Center** in Lynnwood (south of Los Angeles), 295 hospital admissions and ER visits would have been prevented (Figure 4.6). Medi-Cal would have been the primary payer for more than half of these events (156). The next most frequent payer, Medicare, had one-third as many events (51). Nevertheless, Medicare would have enjoyed the largest spending reduction ($716,979), partly because Medi-Cal tends to pay less for hospital care. For example, Medi-Cal spent $9,482 on average for pneumonia admissions those 65 and older, compared with $10,882 for Medicare. Overall, spending at St. Francis would have been reduced by $1,220,595 had federal clean air standards been met (Figure 4.7).

At **Stanford University Hospital**, 30 hospital admissions and ER visits would have been prevented (Figure 4.8), and spending would have been reduced by $534,855 (Figure 4.9). Figure 4.1 shows that fewer events would have been prevented in the San Francisco metro area than in other parts of the state.

At **UC Davis Medical Center** in Sacramento, our final case study, 182 events would have been prevented (Figure. 4.10), and spending would have been reduced by $1,882,412 (Figure 4.11). Medi-Cal was the most frequent payer (81) for these preventable events, while Medicare would have experienced the largest spending reduction ($855,499).

These case studies underscore that health care payers could enjoy substantial reductions in hospital spending from improved air quality. The payers who would benefit the most vary substantially across hospitals and communities.

Figure 4.2
Potentially Prevented Events at Riverside Community Hospital over 2005–2007 Had Federal PM2.5 and Ozone Standards Been Met, by Payer

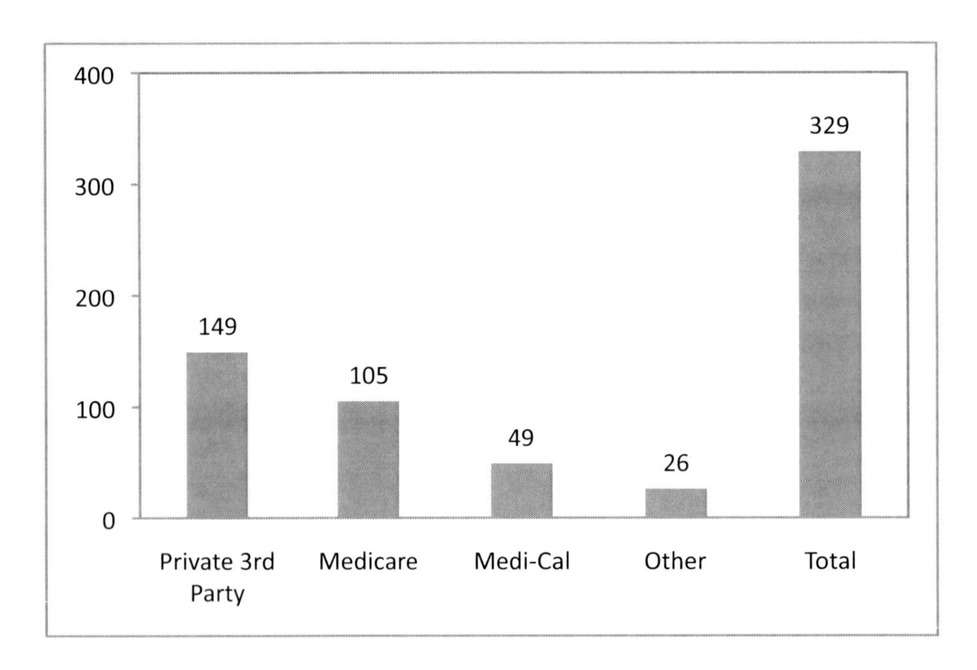

Figure 4.3
Potential Spending Reduction at Riverside Community Hospital over 2005–2007 Had Federal PM2.5 and Ozone Standards Been Met, by Payer

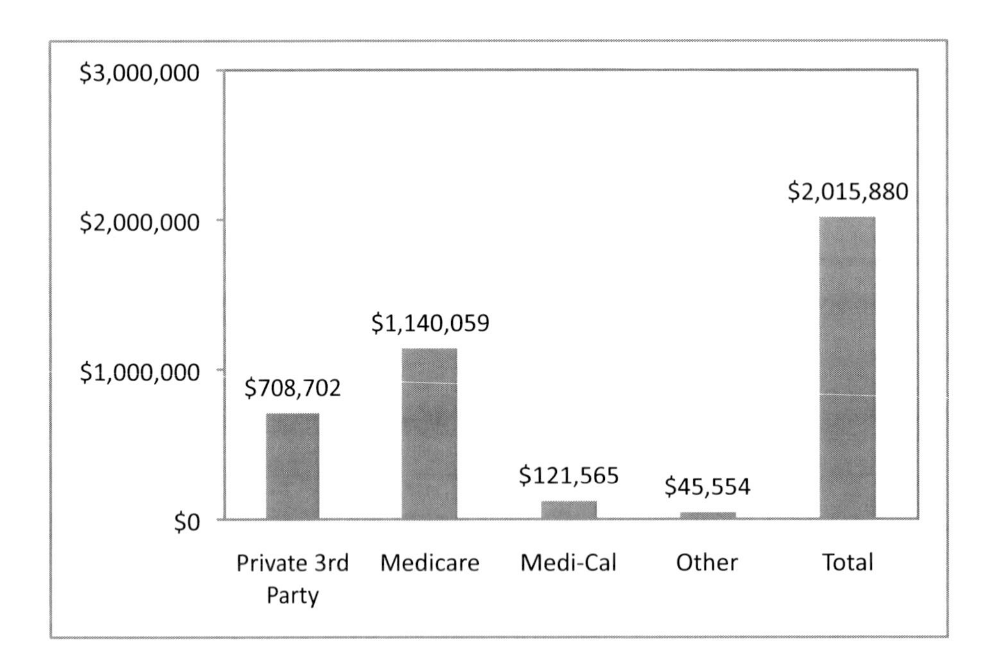

Figure 4.4
Potentially Prevented Events at St. Agnes Medical Center over 2005–2007 Had Federal PM2.5 and Ozone Standards Been Met, by Payer

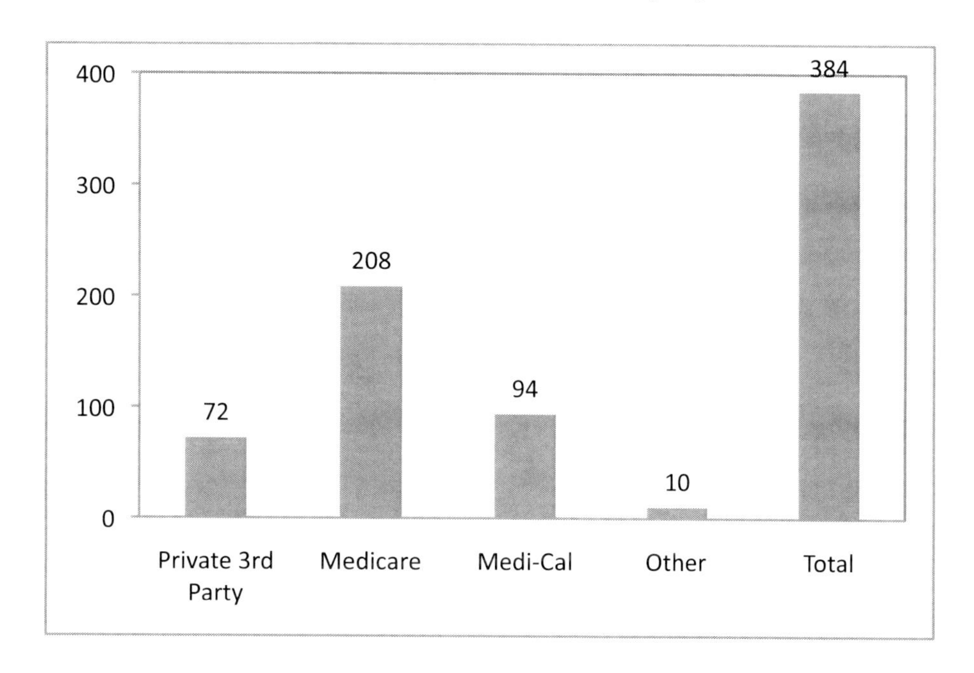

Figure 4.5
Potential Spending Reduction at St. Agnes Medical Center over 2005–2007 Had Federal PM2.5 and Ozone Standards Been Met, by Payer

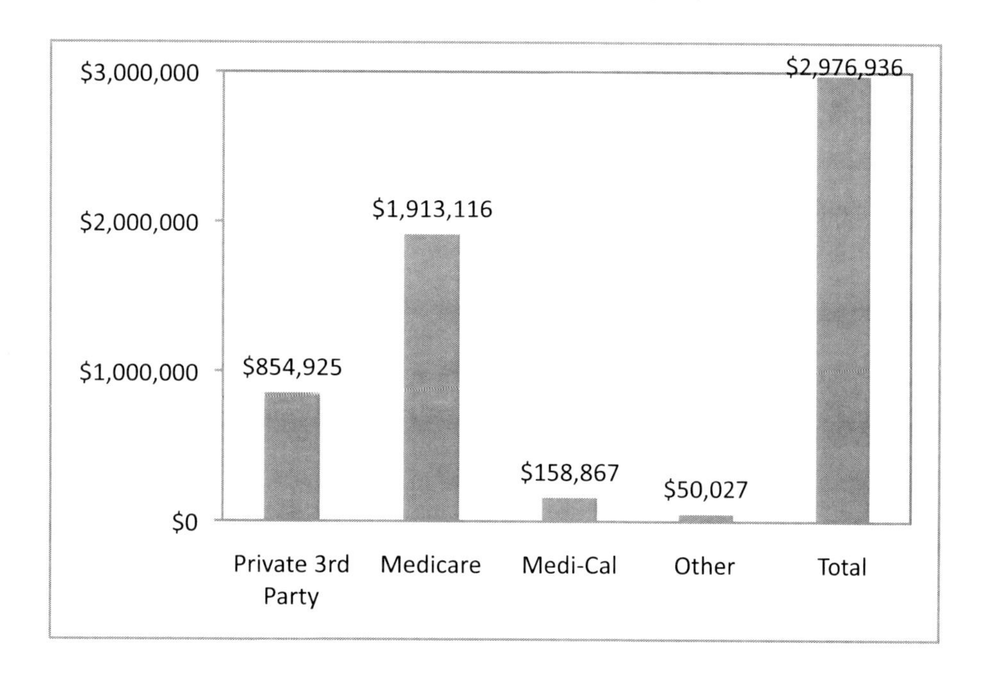

Figure 4.6
Potentially Prevented Events at St. Francis Medical Center over 2005–2007 Had Federal PM2.5 and Ozone Standards Been Met, by Payer

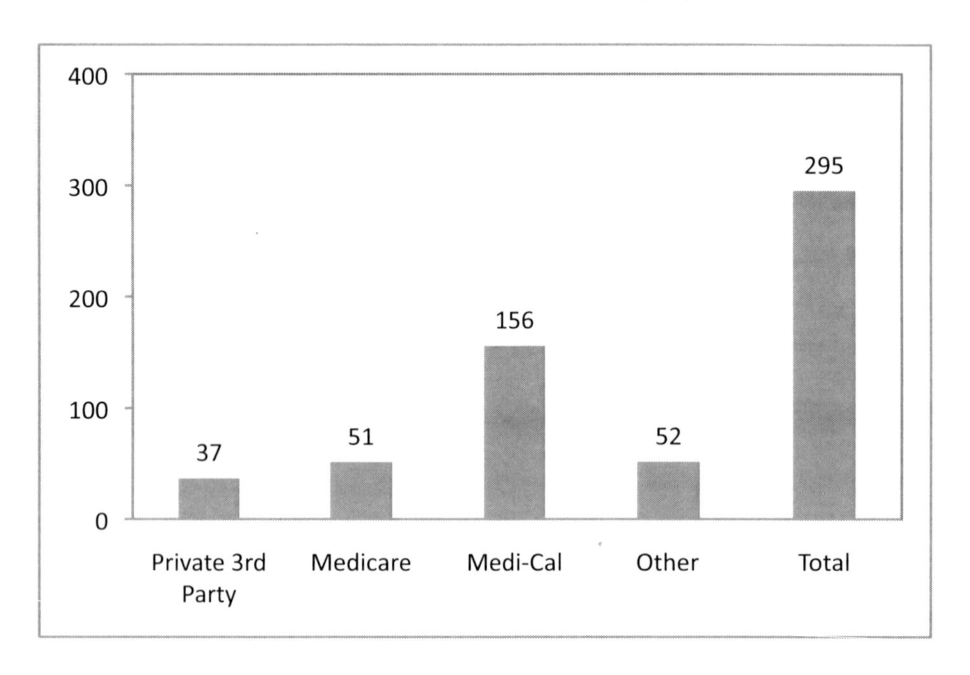

Figure 4.7
Potential Spending Reduction at St. Francis Medical Center over 2005–2007 Had Federal PM2.5 and Ozone Standards Been Met, by Payer

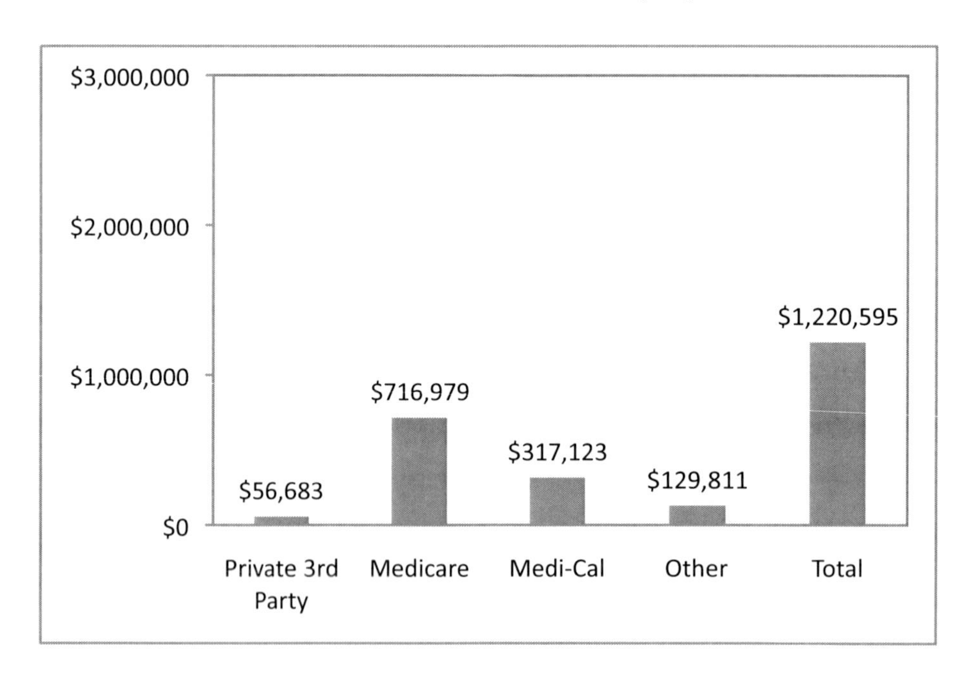

Figure 4.8
Potentially Prevented Events at Stanford University Hospital over 2005–2007 Had Federal PM2.5 and Ozone Standards Been Met, by Payer

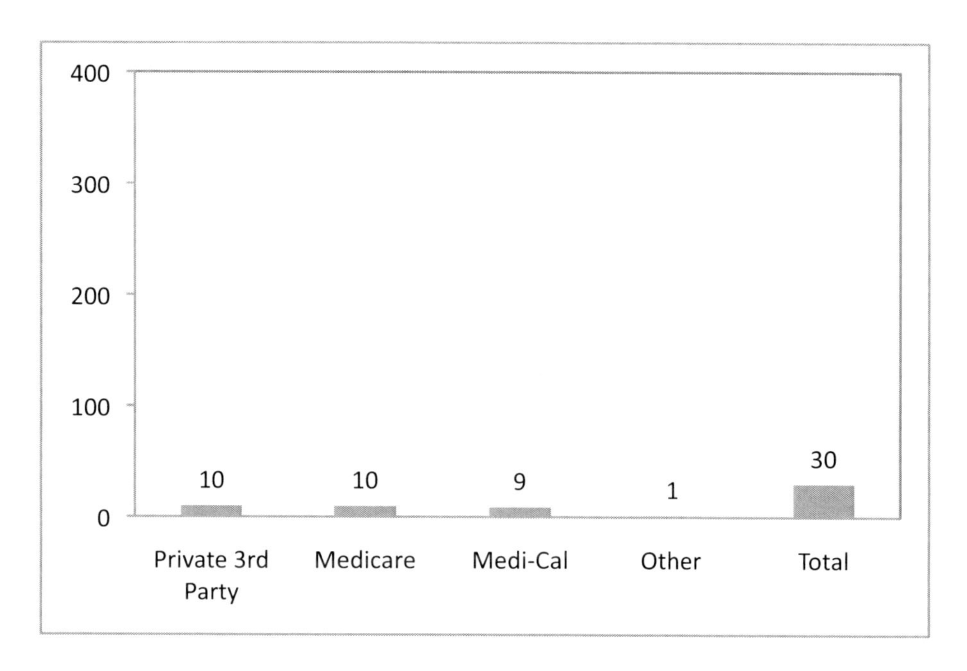

Figure 4.9
Potential Spending Reduction at Stanford University Hospital over 2005–2007 Had Federal PM2.5 and Ozone Standards Been Met, by Payer

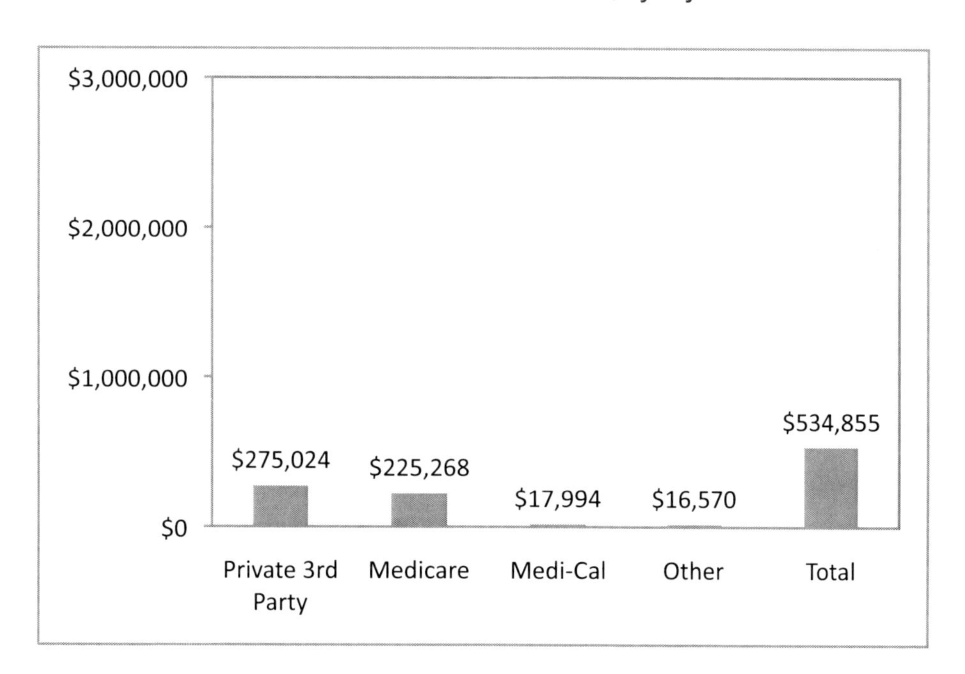

Figure 4.10
Potentially Prevented Events at UC Davis Medical Center over 2005–2007 Had Federal PM2.5 and Ozone Standards Been Met, by Payer

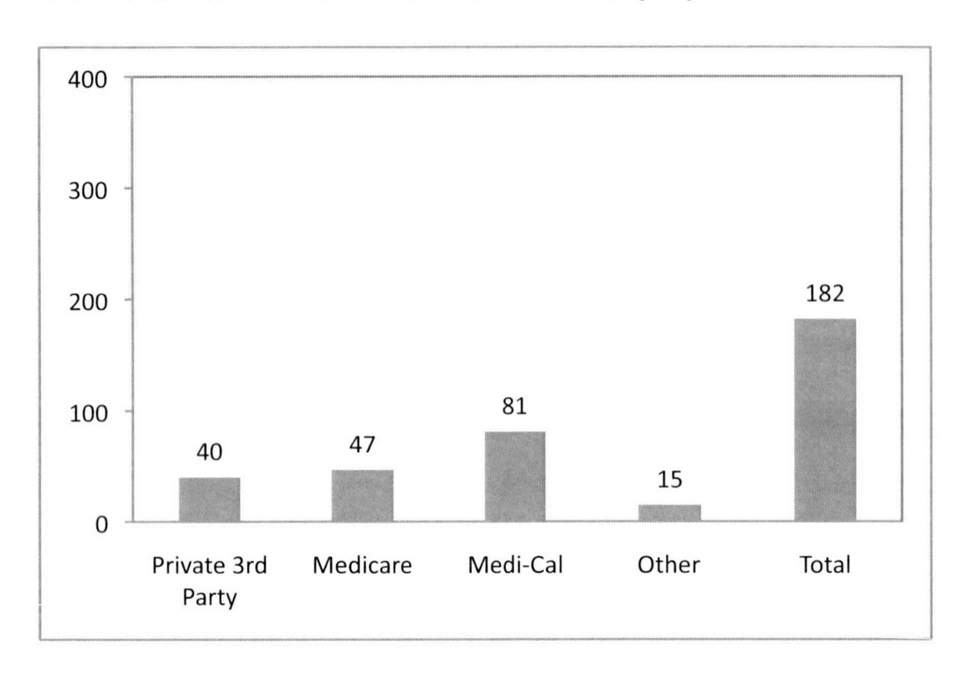

Figure 4.11
Potential Spending Reduction at UC Davis Medical Center over 2005–2007 Had Federal PM2.5 and Ozone Standards Been Met, by Payer

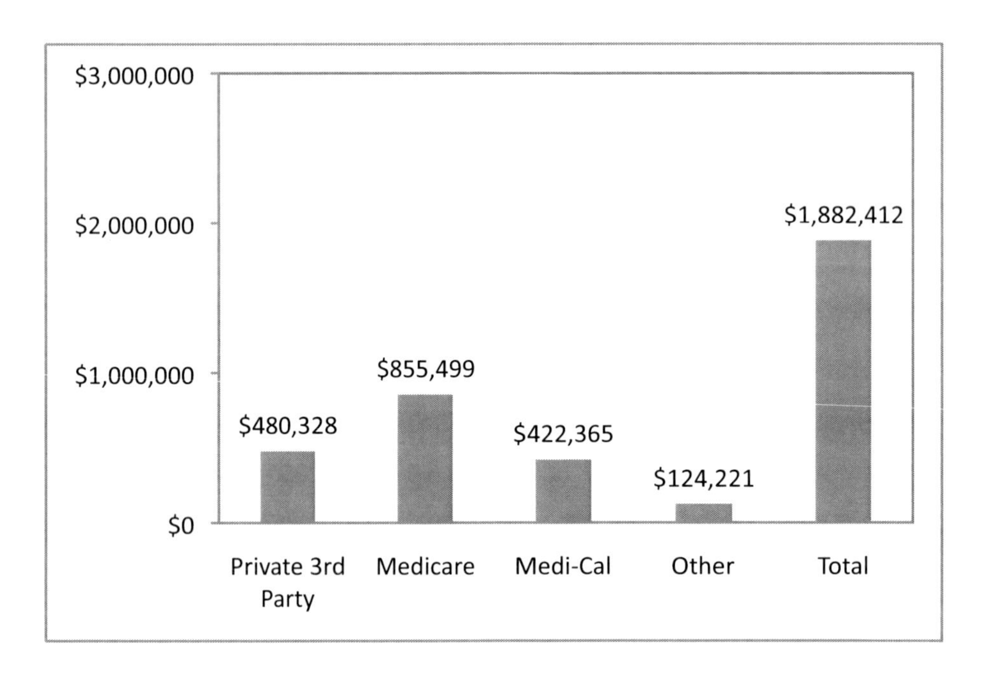

CHAPTER FIVE

Conclusions

This study assessed the financing of pollution-related medical care. In particular, we analyzed hospital admissions and ER visits throughout California from 2005 to 2007.

We found that not meeting federal clean air standards for PM2.5 and ozone caused an estimated 29,808 health events requiring hospital care. Furthermore, public purchasers of health care (including Medicare and Medi-Cal) spent an estimated $131,964,166 on this hospital care, while private third-party purchasers (including managed-care and fee-for-service insurance plans) spent an estimated $55,879,780

These results suggest that the stakeholders of public programs stand to benefit substantially from improved air quality, and that insurance companies and employers may also have sizable stakes in air quality.

It is important to recognize that reduced spending on hospital care could ultimately improve the well-being of others, rather than the bottom lines of insurers. Lower pollution-related health spending could lead to lower insurance premiums, instead of higher profits for insurers, due to marketplace competition. Many employers contribute to employee premiums, and thus might themselves benefit from improved air quality. Yet employees may ultimately be beneficiaries. Workers effectively pay for much of their employer contributions toward health insurance premiums through lower wages (Gruber 1998); thus, lower premiums translate into higher wages.

Nevertheless, there is reason to believe that private insurers and employers would ultimately enjoy some of the benefit from improved air quality. The health insurance industry has been experiencing consolidation in recent years (Robinson 2004). In a market that is not highly competitive, businesses may not fully pass lower costs on to their customers (Tirole 1988). In our setting, health insurance premiums may not fall as much as medical spending does, with higher insurer profits as a result. In the workplace, employee wages sometimes do not offset employer insurance premium contributions fully, due, for example, to long-term union contracts or less formal norms (Sood, Ghosh, et al. 2009). Employers thus benefit from lower insurance premiums.

Results Within Regions and by Detailed Payer Type

The following tables show the main results of our study by county (Table A.1) and by "detailed" payer type (that is, the finest-grained division of payers possible using the available data) (Table A.2). Figures A.1–A.3 show spending by payers throughout the state.

Table A.1
Events, Spending, and Hospital Charges in California over 2005–2007 Caused by Failure to Meet Federal PM2.5 and Ozone Standards, by County, Sorted by Reduction in Spending

County	Events	% of Total Events	Spending	% of Total Spending	Hospital Charges
Los Angeles	12,384	41.54	$83,500,000	43.21	$343,000,000
Orange	2,580	8.66	$18,600,000	9.64	$72,800,000
San Bernardino	2,780	9.33	$15,600,000	8.05	$54,400,000
Riverside	1,999	6.71	$11,600,000	5.98	$43,400,000
Sacramento	1,103	3.70	$9,128,764	4.72	$35,700,000
Fresno	1,977	6.63	$8,761,643	4.53	$28,200,000
Kern	1,348	4.52	$7,136,941	3.69	$26,200,000
Stanislaus	781	2.62	$6,133,071	3.17	$34,700,000
San Joaquin	785	2.63	$5,409,849	2.80	$25,200,000
Tulare	733	2.46	$3,786,777	1.96	$13,700,000
[Unassigned]	467	1.57	$3,609,326	1.87	$13,800,000
San Diego	367	1.23	$2,574,380	1.33	$9,421,399
Santa Clara	169	0.57	$1,814,405	0.94	$6,961,155
Alameda	237	0.79	$1,705,291	0.88	$6,244,745
Merced	351	1.18	$1,525,244	0.79	$7,167,413
Placer	169	0.57	$1,421,900	0.74	$6,081,411
Kings	283	0.95	$1,197,908	0.62	$4,091,745
Contra Costa	138	0.46	$1,177,965	0.61	$4,586,720
Butte	128	0.43	$1,024,691	0.53	$4,515,120
San Francisco	72	0.24	$911,462	0.47	$3,019,130
Madera	229	0.77	$814,746	0.42	$2,374,776
San Mateo	82	0.27	$811,200	0.42	$3,126,933
Yolo	99	0.33	$805,073	0.42	$3,017,442

Table A.1—Continued

County	Events	% of Total Events	Spending	% of Total Spending	Hospital Charges
Ventura	108	0.36	$788,968	0.41	$3,306,287
Solano	74	0.25	$719,897	0.37	$2,816,633
El Dorado	58	0.19	$480,498	0.25	$1,938,666
Sonoma	43	0.15	$388,884	0.20	$1,255,671
Sutter	41	0.14	$358,598	0.19	$1,089,494
Yuba	45	0.15	$326,994	0.17	$1,027,964
Marin	20	0.07	$221,564	0.11	$783,636
Imperial	52	0.17	$212,797	0.11	$722,161
Nevada	30	0.10	$169,532	0.09	$622,713
Napa	10	0.03	$110,532	0.06	$380,906
Shasta	16	0.05	$109,569	0.06	$614,777
Tuolumne	10	0.03	$65,270	0.03	$259,697
Calaveras	8	0.03	$51,832	0.03	$175,430
Santa Cruz	5	0.02	$47,472	0.02	$196,887
Tehama	8	0.03	$44,166	0.02	$186,808
Amador	6	0.02	$37,040	0.02	$143,551
Mariposa	5	0.02	$29,279	0.02	$106,779
Glenn	4	0.01	$26,225	0.01	$107,187
Monterey	3	0.01	$19,744	0.01	$69,819
Colusa	2	< .01	$10,449	< .01	$31,624
Santa Barbara	< 1	< .01	$3,994	< .01	$11,400
San Benito	< 1	< .01	$1,386	< .01	$5,507
Sierra	< 1	< .01	$1,361	< .01	$3,238
Alpine	< 1	< .01	$393	< .01	$1,201
Del Norte	0	0.00	$0	0.00	$0
Humboldt	0	0.00	$0	0.00	$0
Inyo	0	0.00	$0	0.00	$0
Lake	0	0.00	$0	0.00	$0
Lassen	0	0.00	$0	0.00	$0
Mendocino	0	0.00	$0	0.00	$0
Modoc	0	0.00	$0	0.00	$0
Mono	0	0.00	$0	0.00	$0
Plumas	0	0.00	$0	0.00	$0
San Luis Obispo	0	0.00	$0	0.00	$0
Siskiyou	0	0.00	$0	0.00	$0
Trinity	0	0.00	$0	0.00	$0

Table A.2
Events, Spending, and Hospital Charges in California 2005–2007 Caused by Failure to Meet Federal PM2.5 and Ozone Standards, by Detailed Payer Type

Payer	Events	% of Total Events	Spending	% of Total Spending	Hospital Charges
Medicare fee-for-service	6,586	22.09	$72,400,000	37.46	$349,000,000
Medicare managed care	2,661	8.93	$31,200,000	16.13	$114,000,000
Medi-Cal fee-for-service	2,019	6.77	$18,900,000	9.79	$88,400,000
Medi-Cal managed care	865	2.90	$4,427,856	2.29	$21,700,000
Medi-Cal (unidentified)[a]	6,098	20.46	$3,964,343	2.05	$15,900,000
County indigent fee-for-service	321	1.08	$825,248	0.43	$6,783,295
County indigent managed care	14	0.05	$246,719	0.13	$828,838
Third-party fee-for-service	2,744	9.20	$9,073,704	4.70	$22,500,000
Third-party managed care	6,109	20.50	$46,600,000	24.10	$127,000,000
Third-party (unidentified)[a]	176	0.59	$206,076	0.11	$454,889
Other/self-pay	1,955	6.56	$5,087,323	2.63	$19,222,334
Indigent other	261	0.88	$355,685	0.18	$1,697,055

[a] ER records do not identify plan type (i.e., fee-for-service vs. managed care) for Medi-Cal or third-party plans; the label "unidentified" above refers to these records.

Figure A.1
Medicare Spending in California 2005–2007 Caused by Failure to Meet Federal PM2.5 and Ozone Standards, by Congressional District

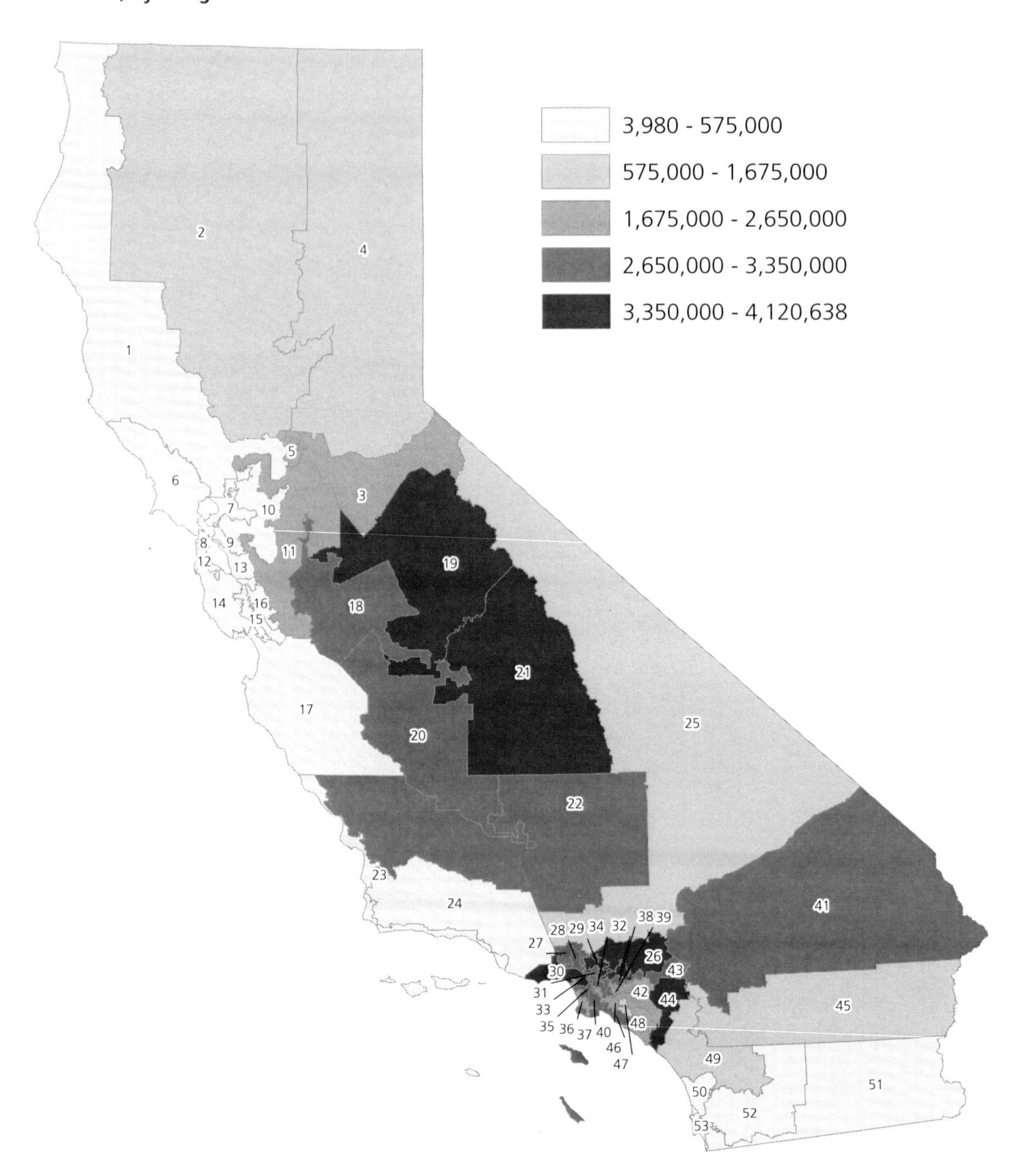

Figure A.2
Medi-Cal Spending in California 2005–2007 Caused by Failure to Meet Federal PM2.5 and Ozone Standards, by Congressional District

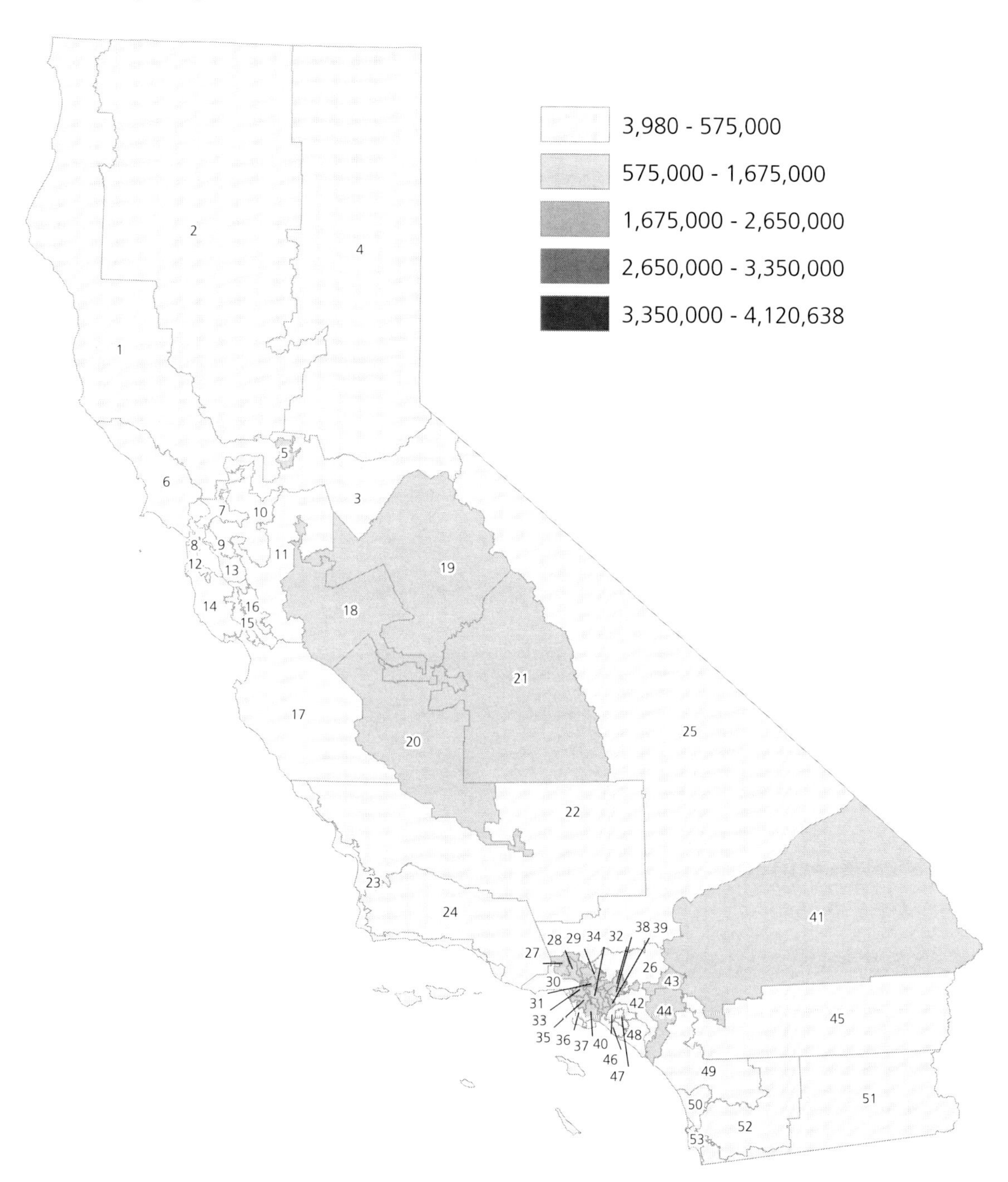

Figure A.3
Private Third-Party Spending in California 2005–2007 Caused by Failure to Meet Federal PM2.5 and Ozone Standards, by County

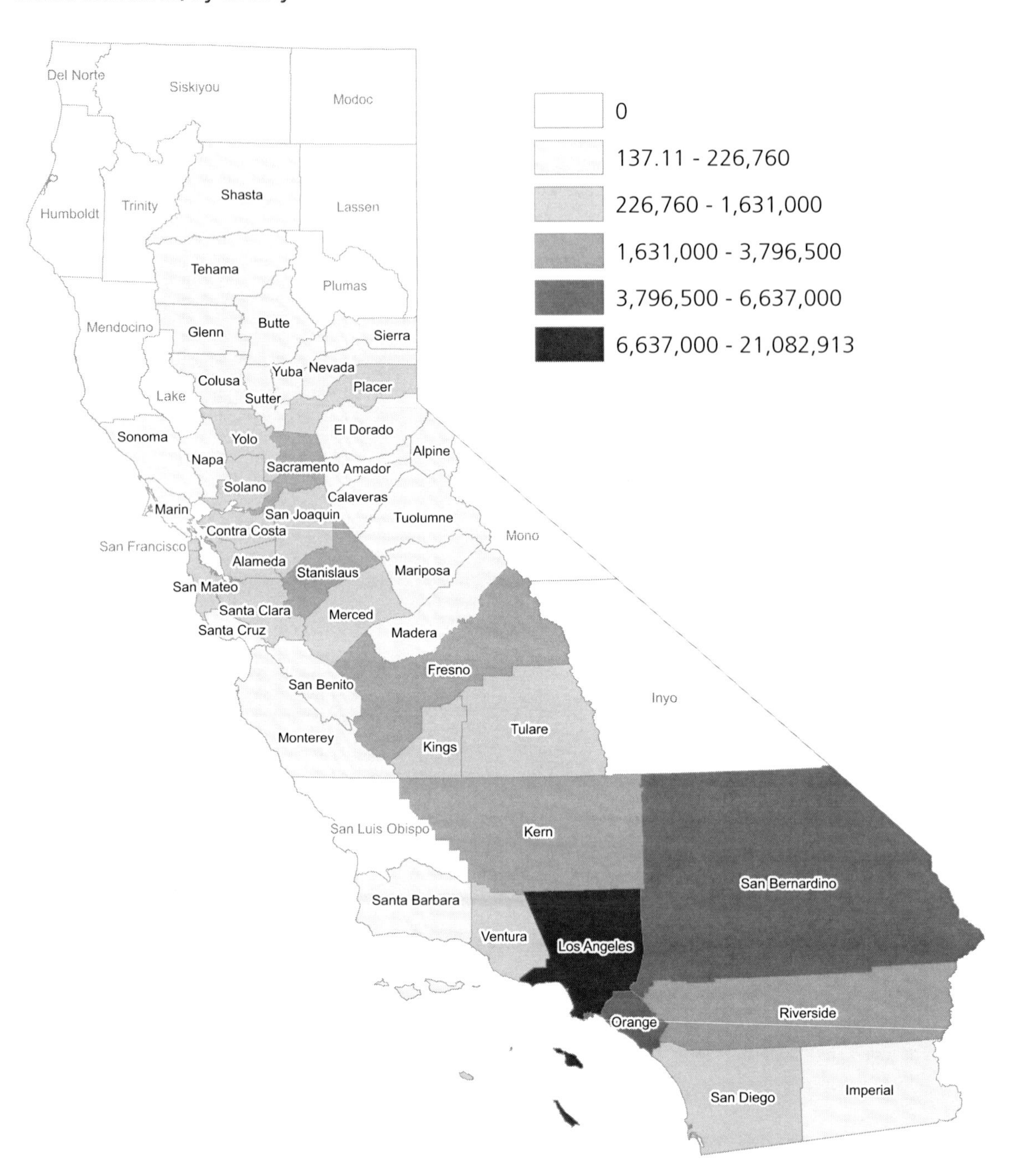

APPENDIX B

Results for Specific Health Plans

The Patient Discharge Data Files identify certain health plans as the actual or expected primary payer for care. In particular, the files identify health care service plans (including health maintenance organizations) licensed under California's Knox-Keene Healthcare Service Plan Act, as well as Medi-Cal County Organized Health Systems.

We determined events prevented and spending reductions for these specific health plans. Table B.1 reports the results.

It is important to recognize that these results do not include ER visits. The Emergency Department and Ambulatory Surgery Data Files do not identify plans. The identified plans account for 33.3% of all potentially prevented events and 28.8% of the overall potential spending reduction for hospital admissions in Table 4.1.

Table B.1
Events, Spending, and Hospital Charges for Hospital Admissions in California over 2005–2007 Caused by Failure to Meet Federal PM2.5 and Ozone Standards, by Specific Health Plan, Ordered by Reduction in Spending

Payer Plan Name	Events	Spending	Hospital Charges
Kaiser Foundation Health Plan, Inc.	1,827	$30,000,000	$86,800,000
Secure Horizons	899	$10,400,000	$38,400,000
Blue Cross of California	538	$5,006,344	$17,700,000
Blue Shield of California	308	$3,769,777	$12,700,000
Health Net of California, Inc.	368	$3,717,733	$13,500,000
SCAN Health Plan	245	$2,094,761	$8,444,284
Aetna Health Plans of California, Inc.	115	$1,227,700	$4,077,754
Other HMO	83	$897,481	$3,596,255
Universal Care	32	$662,573	$946,750
Cigna HealthCare of California, Inc.	44	$586,641	$1,830,143
Inland Empire Health Plan (IEHP)	127	$586,306	$2,519,441
Cal Optima (Orange County)	74	$579,153	$2,957,885
UHP Healthcare	43	$555,919	$1,794,101
Inter Valley Health Plan	41	$465,577	$1,836,608
Managed Health Network	38	$411,434	$1,393,245
Caloptima (Orange County)	47	$344,934	$1,618,788
Care 1st Health Plan	47	$322,207	$1,198,414

Table B.1—Continued

Payer Plan Name	Events	Spending	Hospital Charges
American Family Care	68	$316,143	$1,410,722
Community Health Plan (County of Los Angeles)	41	$264,488	$1,035,614
Western Health Advantage	15	$261,476	$932,057
Kern Health Systems Inc	35	$245,151	$798,602
CareMore Insurance Services, Inc.	36	$221,528	$1,214,582
LA Care Health Plan	22	$181,868	$460,911
UHC Healthcare	10	$132,273	$400,672
Primecare Medical Network, Inc.	5	$88,580	$246,298
The Health Plan of San Joaquin	15	$85,159	$435,905
Heritage Provider Network, Inc.	8	$58,915	$217,573
AET Health Care Plan Of California	5	$52,459	$277,952
Great-West Healthcare of California, Inc.	4	$50,678	$161,539
Brown and Toland Medical Group	1	$44,183	$94,910
San Francisco Health Plan	1	$43,205	$39,400
Solano Partnership Health Plan (Solano County)	6	$42,786	$263,672
Chinese Community Health Plan	2	$37,094	$74,214
Community Health Group	8	$36,748	$152,808
Medcore HP	1	$36,070	$40,851
Health Plan Of San Mateo	3	$34,594	$188,991
Scripps Clinic Health Plan Services, Inc.	5	$27,867	$131,977
Sharp Health Plan	3	$25,916	$69,137
Alameda Alliance for Health	4	$22,331	$102,890
Health Plan of San Mateo (San Mateo County)	2	$19,297	$84,279
Santa Clara Valley Medical Center	1	$16,025	$36,642
Contra Costa Health Plan	2	$13,984	$66,947
PacifiCare Behaviorial Health of California	1	$12,824	$31,452
Santa Clara Family Health Plan	1	$4,470	$35,121
Tower Health Service	1	$4,221	$28,562
Cigna Behaviorial Health of California	< 1	$3,925	$10,942
Central Coast Alliance For Health (Santa Cruz County/Montery County)	< 1	$2,578	$26,261
Ventura County Health Care Plan	< 1	$2,121	$7,275
On Lok Senior Health Services	< 1	$1,909	$7,143
Central Coast Alliance for Health (Santa Cruz County)	< 1	$1,782	$7,582
Avante Behavioral Health Plan	< 1	$1,259	$1,668
Central Health Plan	< 1	$1,223	$15,375

Table B.1—Continued

Payer Plan Name	Events	Spending	Hospital Charges
[Unknown]	< 1	$527	$2,137
Santa Barbara Health Authority (Santa Barbara County)	< 1	$451	$2,671
Health Plan of the Redwoods	< 1	$285	$2,128
Vista Behaviorial Health Plan	< 1	$67	$272
Lifeguard, Inc.	< 1	$64	$166
ProMed Health Care Administrators	< 1	$0	$612

Additional Details for Analysis Step 2

We used hospital care data to determine the number of endpoint events that would have been prevented under our air quality improvement scenario. In some cases, relevant patient characteristics were missing or redacted in the data.

When possible, each record with an unassigned patient age category (7.89% of discharge records, 4.07% of ER visit records) was assigned a category using a Monte Carlo approach based on partitioned diagnosis group,[1] patient zip code, and detailed payer type. Records for which age could not be predicted in this manner because no matching records existed—2.36% of discharge records and 2.32% of ER visit records—were assigned an age category based only on endpoint and payer. Less than .001% of records remained unassigned after this exercise; these records were excluded from the analysis.

Records with missing quarter of care (3.60% of discharge records, 2.72% of ER visit records) were assigned a quarter using a Monte Carlo approach based on diagnosis group, patient zip code, and age category. Records for which a quarter could not be predicted in this manner—2.00% of discharge records and 2.18% of ER visit records—were assigned a quarter based only on endpoint and age. No records remained unassigned after this exercise.

We also considered patient race and ethnicity. This field was much more likely to be redacted or missing the data file than quarter or age: 29.66% of discharge records and 21.14% of ER visit records were missing race. When possible, each record with missing patient race/ethnicity was assigned to a category using a Monte Carlo approach based on diagnosis group, patient zip code, and detailed payer type. Records for which a race/ethnicity category could not be predicted in this manner—3.46% of discharge records and 2.94% of ER visit records—were assigned to a category based only on endpoint and payer. No records remained unassigned after this exercise.

Finally, records with missing or incomplete patient zip code (2.70% of discharge records and 3.57% of ER visit records) were excluded from the analysis.

---

[1] In a sensitivity analysis in Appendix E, we consider alternative definitions of endpoints based on International Classification of Disease codes for principal diagnoses. Where alternative endpoint definitions partially overlap with respect to diagnoses, a missing patient characteristic could be assigned based on the distribution of the characteristic for either endpoint definition. To deal with this ambiguity, we partitioned the diagnoses for such into unique and common groups and used the resulting partitioned groups to assign missing characteristics.

Payer Category Assignment

Patient discharge records include information on the expected source of payment for hospital care and the type of insurance coverage when applicable. ER records include a single field describing the source of payment for care. The quarterly financial reports submitted by hospitals decompose gross and net revenues by a payer category. Tables D.1 and D.2 show the crosswalk assignments we made between patient records—both discharge and ER visits—and hospital financial report payer categories. Note that the term "traditional"—which is used in the discharge data, the ER visit data, and the hospital financial report data described below—is equivalent to what we have referred to as "fee-for-service."

Table D.1
Assignments of Inpatient Discharge Payer to Hospital Financial Report Payer Categories

Patient Payer Category	Patient Payer Type	Hospital Financial Report Payer Category
Medicare	Traditional	Medicare—traditional
Medicare	Managed care	Medicare—managed care
Medi-Cal	Traditional	Medi-Cal—traditional
Medi-Cal	Managed care	Medi-Cal—managed care
County indigent program	Traditional	County indigent program—traditional
County indigent program	Managed care	County indigent program—managed care
Private coverage, workers' compensation, or other government	Traditional	Other third parties—traditional
Private coverage, workers' compensation, or other government	Managed care	Other third parties—managed care
Self-pay or other payer	Any	Other payers
Other indigent	Any	Other indigent
Not assigned	Any	Not assigned

Table D.2
Assignments of ER Visit Payer to Hospital Financial Report Payer Categories

ER Visit Expected Source of Payment	Hospital Financial Report Payer Category
Medicare Part A or B	Medicare—traditional
HMO Medicare Risk	Medicare—managed care
Medicaid	Medi-Cal—unidentified[a]
Other Nonfederal Programs	County indigent program—traditional
PPO, POS, EPO, Blue Cross/Blue Shield, commercial insurance, Disability, other federal program, Title V, workers' compensation	Other third parties—traditional
Health Maintenance Organization	Other third parties—managed care
Automobile Medical, CHAMPUS/TRICARE, Veterans Affairs Plan	Other third parties—unidentified[a]
Self-pay	Other payers
Other	Other indigent
Invalid/blank	Not assigned

[a] ER records do not identify plan type (i.e., traditional versus managed care) for Medi-Cal or third-party plans; the label "unidentified" above refers to these records. Discounts for these categories are calculated as the weighted average of traditional and managed care discounts for that payer, using total outpatient volume as a weighting factor.

References

Abt Associates (2008). Environmental Benefits Mapping and Analysis Program (BenMAP), User's Manual.

Bell, M. L., R. D. Peng, et al. (2006). "The exposure-response curve for ozone and risk of mortality and the adequacy of current ozone regulations." *Environmental Health Perspectives* 114(4): 532–536.

Burnett, R. T., M. Smith-Doiron, et al. (2001). "Association between ozone and hospitalization for acute respiratory diseases in children less than 2 years of age." *American Journal of Epidemiology* 153(5): 444–452.

California Air Resources Board (2002a). "Air Board Passes Stronger Particulate Matter Standards." Press release, June 20, 2002. As of December 16, 2009:
http://www.arb.ca.gov/newsrel/nr062002.htm

California Air Resources Board (2002b). "Public Hearing to Consider Amendments to the Ambient Air Quality Standards for Particulate Matter and Sulfates, May 3, 2002." As of December 16, 2009:
http://www.arb.ca.gov/research/aaqs/std-rs/pm-final/pm-final.htm

California Air Resources Board (2005a). "California Adopts New Ozone Standard Children's Health Focus of New Requirement." Press release, April 28, 2005. As of December:
http://www.arb.ca.gov/newsrel/nr042805.htm

California Air Resources Board (2005b). "Review of the California Ambient Air Quality Standard for Ozone." Air Quality Advisory Committee Meeting, January 11–12, 2005. As of December 16, 2009:
http://www.arb.ca.gov/research/aaqs/ozone-rs/aqac/aqac.htm

California Air Resources Board (2006). "Emission Reduction Plan for Ports and Goods Movement." As of December 16, 2009:
http://www.arb.ca.gov/planning/gmerp/gmerp.htm

California Air Resources Board (2009). U.S. EPA Submittal—ARB Recommended Area Designations for the 2008 Federal 8-Hour Ozone Standard, Enclosure 1: Recommended California Nonattainment Areas for the Federal 8-Hour Ozone Standard. As of December 16, 2009:
http://www.arb.ca.gov/desig/8-houroz/8-houroz.htm

California Office of Statewide Health Planning and Development (2009a). "Emergency Department and Ambulatory Surgery Data File."

California Office of Statewide Health Planning and Development (2009b). "Hospital Quarterly Financial and Utilization Data File." As of December 16, 2009:
http://www.oshpd.ca.gov/hid/Products/Hospitals/QuatrlyFinanData/CmpleteData/default.asp

California Office of Statewide Health Planning and Development (2009c). "Patient Discharge Data File." Retrieved September 24, 2009, from:
http://www.oshpd.ca.gov/HID/Products/PatDischargeData/PublicDataSet/index.html

Centers for Disease Control (2009). "CDC Vaccine Price List." As of December 16, 2009:
http://www.cdc.gov/vaccines/programs/vfc/cdc-vac-price-list.htm#adflu

Chestnut, L. G., M. A. Thayer, et al. (2006). "The economic value of preventing respiratory and cardiovascular hospitalizations." *Contemporary Economic Policy* 24(1): 127–143.

EPA—*See* U.S. Environmental Protection Agency.

Fiore, A. M., D. J. Jacob, et al. (2002). "Background ozone over the United States in summer: Origin, trend, and contribution to pollution episodes." *J. Geophys. Res.* 107.

Fuchs, V. R., and S. R. Frank (2002). "Air pollution and medical care use by older Americans: A cross-area analysis." *Health Affairs* (Project Hope) 21(6): 207–214.

Gruber, J. (1998). *Health Insurance and the Labor Market.* Cambridge, MA: National Bureau of Economic Research.

Hall, J. V., V. Brajer, and F. W. Lurmann (2008a). *The Benefits of Meeting Federal Clean Air Standards in the South Coast and San Joaquin Valley Air Basins.* Fullerton, CA: Institute for Economic and Environmental Studies, California State University, Fullerton. As of January 4, 2010:
http://business.fullerton.edu/centers/iees/reports/Benefits_of_Meeting_Clean_Air_Standards_11-13-08.pdf

Hall, J. V., V. Brajer, and F. W. Lurmann (2008b). "Measuring the gains from improved air quality in the San Joaquin Valley." *Journal of Environmental Management* 88(4): 1003–1015.

Hartman, M., A. Martin, et al. (2009). "National health spending in 2007: Slower drug spending contributes to lowest rate of overall growth since 1998." *Health Affairs* 28(1): 246–261.

Holgate, S. T., H. S. Koren, et al. (1999). *Air Pollution and Health*, Academic Press.

Ito, K. (2003). Associations of Particulate Matter Components with Daily Mortality and Morbidity in Detroit, Michigan. *Revised Analyses of Time-Series Studies of Air Pollution and Health.* Boston, MA, Health Effects Institute.

Jaffe, D. H., M. E. Singer, et al. (2003). "Air pollution and emergency department visits for asthma among Ohio Medicaid recipients, 1991–1996." *Environ Research* 91(1): 21–28.

Jerrett, M., R. T. Burnett, et al. (2009). "Long-term ozone exposure and mortality." *New England Journal of Medicine* 360(11): 1085–1095.

Luft, H. S., D. W. Garnick, et al. (1990). "Does quality influence choice of hospital?" *Journal of the American Medical Association* 263(21): 2899–2906.

Moolgavkar, S. H. (2000a). "Air pollution and hospital admissions for chronic obstructive pulmonary disease in three metropolitan areas in the United States." *Inhalation Toxicology* 12 Suppl 4: 75–90.

Moolgavkar, S. H. (2000b). "Air pollution and hospital admissions for diseases of the circulatory system in three U.S. metropolitan areas." *Journal of the Air and Waste Management Association* 50(7): 1199–1206.

Moolgavkar, S. H. (2003). "Air Pollution and Daily Deaths and Hospital Admissions in Los Angeles and Cook Counties." In *Revised Analyses of Time-Series Studies of Air Pollution and Health.* Boston, MA: Health Effects Institute.

Moolgavkar, S. H., E. G. Luebeck, et al. (1997). "Air pollution and hospital admissions for respiratory causes in Minneapolis-St. Paul and Birmingham." *Epidemiology* (Cambridge, MA) 8(4): 364–370.

National Heart, Lung, and Blood Institute (undated). "Take the First Step to Breathing Better. Learn More About COPD." As of December 17, 2009:
http://www.nhlbi.nih.gov/health/public/lung/copd/

Neidell, M. J. (2004). "Air pollution, health, and socio-economic status: The effect of outdoor air quality on childhood asthma." *Journal of Health Economics* 23(6): 1209–1236.

Norris, G., S. N. YoungPong, et al. (1999). "An association between fine particles and asthma emergency department visits for children in Seattle." *Environmental Health Perspectives* 107(6): 489–493.

O'Neill, M. S., M. Jerrett, et al. (2003). "Health, wealth, and air pollution: Advancing theory and methods." *Environmental Health Perspectives* 111(16): 1861–1870.

Peel, J. L., P. E. Tolbert, et al. (2005). "Ambient air pollution and respiratory emergency department visits." *Epidemiology* (Cambridge, MA) 16(2): 164–174.

Robinson, J. C. (2004). "Consolidation and the transformation of competition in health insurance." *Health Affairs* 23(6): 11–24.

Schwartz, J. (1994). "PM10, ozone, and hospital admissions for the elderly in Minneapolis-St. Paul, Minnesota." *Archives of Environmental Health* 49(5): 366–374.

Sheppard, L. (2003). "Ambient Air Pollution and Non-elderly Asthma Hospital Admissions in Seattle, Washington, 1987–1994." In *Revised Analyses of Time-Series Studies of Air Pollution and Health.* Boston, MA: Health Effects Institute.

Smith, J. P. (1999). "Healthy bodies and thick wallets: The dual relation between health and economic status." *Journal of Economic Perspectives* 13(2): 145–166.

Sood, N., A. Ghosh, et al. (2009). "Employer-sponsored insurance, health care cost growth, and the economic performance of U.S. industries." *Health Services Research* 44(5 Pt 1): 1449–1464.

Thurston, G. D., and K. Ito (1999). "Epidemiological studies of ozone exposure effects." In *Air Pollution and Health.* S. T. Holgate, H.S. Koren, et al., eds. San Diego, CA: Academic Press.

Tirole, J. (1988). *The Theory of Industrial Organization.* Cambridge, MA: MIT Press.

U.S. Census Bureau (2009). "American Community Survey." As of December 16, 2009:
http://www.census.gov/acs/

U.S. Environmental Protection Agency (1999a). *The Benefits and Costs of the Clean Air Act: 1990 to 2010.* As of December 16, 2009:
http://www.epa.gov/air/sect812/prospective1.html

U.S. Environmental Protection Agency (1999b). "The Cost of Illness Handbook." As of December 16, 2009:
http://www.epa.gov/oppt/coi/index.html

U.S. Environmental Protection Agency (2003). *Benefits and Costs of the Clean Air Act 1990–2020: Revised Analytical Plan for EPA's Second Prospective Analysis.* As of December 16, 2009:
http://www.epa.gov/air/sect812/blueprint.html

U.S. Environmental Protection Agency (2004). *Advisory on Plans for Health Effects Analysis in the Analytical Plan for EPA's Second Prospective Analysis—Benefits and Costs of the Clean Air Act, 1990–2020.*

U.S. Environmental Protection Agency (2005). *Regulatory Impact Analysis for the Final Clean Air Interstate Rule.* As of December 16, 2009:
http://www.epa.gov/cair/pdfs/finaltech08.pdf

U.S. Environmental Protection Agency (2006a). *Air Quality Criteria for Ozone and Related Photochemical Oxidants (2006 Final).*

U.S. Environmental Protection Agency (2006b). *Regulatory Impact Analysis of 2006 NAAQS for Particle Pollution.*

U.S. Environmental Protection Agency (2007). "Area Designations for the Revised 24-Hour Fine Particle National Ambient Air Quality Standards," memorandum from Robert J. Meyers, Acting Assistant Administrator, Office of Air and Radiation, to Regional Administrators, Regions I–X.

U.S. Environmental Protection Agency (2008a). "Area Designations for 2006 24-Hour Fine Particle (PM2.5) Standards." As of December 16, 2009:
http://www.epa.gov/pmdesignations/2006standards/index.htm

U.S. Environmental Protection Agency (2008b). "Area Designations for the 2008 Revised Ozone National Ambient Air Quality Standards," memorandum from Robert J. Meyers, Principal Deputy Assistant Administrator, Office of Air and Radiation to Regional Administrators, Regions I–X.

U.S. Environmental Protection Agency (2008c). *Final Ozone NAAQS Regulatory Impact Analysis.*

U.S. Environmental Protection Agency (2008d). "Integrated Science Assessment for Particulate Matter (External Review Draft)." As of December 16, 2009:
http://www.epa.gov/EPA-AIR/2008/December/Day-19/a30197.htm

U.S. Environmental Protection Agency (2008e). Letter to Governor Arnold Schwarzenegger, August 18, 2008.

U.S. Environmental Protection Agency (2009a). "Criteria Pollutant Area Summary Report." As of December 16, 2009:
http://www.epa.gov/oar/oaqps/greenbk/ancl2.html

U.S. Environmental Protection Agency (2009b). "National Ambient Air Quality Standards." As of December 16, 2009:
http://epa.gov/air/criteria.html

Wilson, A. M., C. P. Wake, et al. (2005). "Air pollution, weather, and respiratory emergency room visits in two northern New England cities: An ecological time-series study." *Environmental Research* 97(3): 312–321.

ZIPList5 (2009). (Commercial database.) The Woodlands, TX: ZipInfo.